Denkschriften der Schweizerischen Naturforschenden Gesellschaft
Mémoires de la Société Helvétique des Sciences Naturelles
Band/Vol. 96

Augusto Gansser

Geologisches Institut der
Eidgenössischen Technischen Hochschule
und der Universität, Zürich

Geology of the Bhutan Himalaya

1983
Birkhäuser Verlag
Basel · Boston · Stuttgart

Edited by the commission for memorandums of the Swiss Society for Nature Research with subventions of the Swiss Federation and published with the support of the Swiss National Science Foundation.

Author's address:
Prof. Dr. Augusto Gansser
Eidgenössische Technische Hochschule
Geologisches Institut
Sonneggstrasse 5
CH-8092 Zürich (Switzerland)

Illustration of the front cover:
The Takhtsang Monastery, the type locality of the Takhtsang Gneiss, the predominant rock type of the crystalline thrust sheet.
(Phot. A. Gansser)

Illustration of the back cover:
Royal Bhutanese seal.

Illustration of the title page:
The Gaza Dzong with the Kang Bum mountain, exposing the eastwards rising Lingshi synclinorium with Tethyan sediments. NW Bhutan.

Library of Congress Cataloging in Publication Data

Gansser, Augusto, 1910–
Geology of the Bhutan Himalaya.
(Denkschriften der Schweizerischen Naturforschenden Gesellschaft ; Bd. 96 = Mémoires de la Société helvétique des sciences naturelles ; v. 96)
Bibliography: p.
1. Geology Himalaya Mountains. 2. Geology Bhutan.
I. Title. II. Series: Denkschriften der Schweizerischen Naturforschenden Gesellschaft ; Bd. 96.
QE319.H5G29 1983 555.49'8 83-10028
ISBN 3-7643-1371-4

CIP-Kurztitelaufnahme der Deutschen Bibliothek

Gansser, Augusto:
Geology of the Bhutan Himalaya / Augusto Gansser. - Basel ; Boston ; Stuttgart : Birkhäuser, 1983.
(Denkschriften der Schweizerischen Naturforschenden Gesellschaft ; Bd. 96)
ISBN 3-7643-1371-4
NE: Schweizerische Naturforschende Gesellschaft: Denkschriften der Schweizerischen ...

Layout: Sonja Benz, Stein/AG
Printed in Switzerland
ISBN 3-7643-1371-4

In Memory of H.M. King Jigme Dorje Wangchuck

Contents

A. Introduction

Of the 2500 km of Himalayan range, the Bhutan Himalaya is probably one of the least known sections. An isolated country, since its early history, Bhutan is only beginning to open its borders to a very limited and strictly localized tourism. Situated in the eastern Himalaya it grows a luxuriant vegetation which covers in some areas practically vertical cliffs and merges with the snow line. Only the border-regions towards Tibet are sufficiently barren to reveal the regional geology, but the frequent snow falls in this highest part of the range again mask much of the geological outcrops.

A.1. Previous work

Godwin-Austen, Mallet, Pilgrim and *Hayden* were the only known geologists to have published their investigations over parts of Bhutan prior to our own research work and this was over 60 years ago! The one exception is the more comprehensive work on the Baxa Duars published in 1941 by A. *Lahiri* which only dealt with the very southwestern border of the Bhutan Himalaya area.

H.H. *Godwin-Austen* (1868) and F.R. *Mallet* (1875) were the first to visit the southwestern foothills of Bhutan during investigations in connection with mineral resources of the Darjeeling area. *Mallet* observed that the Siwaliks between Sikkim and Bhutan were missing and his accurate maps indicate a relief thrust of the Lower Himalaya into a preexisting erosional gap in the Siwalik molasse (*Heim* and *Gansser*, 1939; *Gansser*, 1964). His main contribution to Bhutan geology was the recognition of the Baxa series which he named after Baxa Fort in the western Duars (SW Bhutan).* It outcrops intermittently between the Siwalik Molasse and the older foothill rocks. A most enigmatic «outcrop of basement rocks» was discovered by *Mallet* (1875) in the Raidak (Wong Chu) river south of the Siwaliks, which will be discussed later (p. 16).

30 years after *Mallet*, *Pilgrim* (1906) traversed the foothills of eastern Bhutan, partly in connection with coal possibilities (fig. 1), and in 1907 the famous H.H. *Hayden* in his inspiring publication on central southern Tibet (Tsang and U) suggested the possibility of a continuation of Tethyan sediments into the Chomolhari area in NW Bhutan which we were able to confirm in 1963. In 1941 A. *Lahiri* published a detailed account on the Baxa Duars, adding some new information to the observations of *Mallet* and *Pilgrim*. He clearly separated a lower Sinchu-La «stage» from a higher Baxa «stage», while placing both into the Dalings. Within the Sinchu-La argillaceous horizons he notes that the metamorphic grade increases to the N (up the sequence), terminating in Darjeeling type biotite gneisses. His conclusions are somewhat difficult to follow, perhaps due to the fact that some samples were not in situ. A valuable correlation with *Mallet* and *Pilgrim's* observations is given with comparison of other localities within the eastern and western Himalaya (fig. 2).

The Geological Survey of India began regional investigations in the foothill zone of Bhutan after 1960, including some traverses into central Bhutan. An excellent account of this work was published by *Nautiyal et al.* (1964) for the International Geological Congress in New Delhi, shortly preceded by the publication of my book «Geology of the Himalayas», which also contains a preliminary account of the geology of Bhutan based on our first reconnaissance in 1963. In both these nearly «simultaneous» publications new names were introduced which do not always coincide and the matter of priority arises! The survey geologists had already recognized widespread crystalline rocks covering a great part of Bhutan. Of special interest is a tectonic succession, including the Thethyan sediments of NW Bhutan, with 8 major thrust zones. A rough sketch map is included covering central and southern Bhutan, however, without showing the northern areas with the High Himalaya range.

In spite of detailed investigations by *Nautiyal et al.* (1964) the stratigraphy and correlations of the formations in the foothill structures are not convincing. The existence of a marine and terrestrial Damuda facies separated by the «Phuntsholing series» is open to doubt. This latter subdivision is placed between the Dalings and the Baxa «series» and regarded as an equivalent of the purple slates underlying the Krols. With this correlation the Baxas, which where reduced from the original section of *Mallet* and *Lahiri*, should correlate with the Blaini-Krol-Tal group and consequently be of Permo/Triassic age. Since the Baxa «series» are, apart from stromatolites (*Acharyya*, 1974) nonfossiliferous, their correlation with the Permo/Triassic Krol series is doubtful and an older, probably late Pre to Eocambrian age, such as already suggested by *Mallet* and *Lahiri* seems more likely.

*(Baxa and Buxa are both in use. *Mallet* (1875) speaks of the Baxa series, but has Buxa on his map. *Lahiri* (1941) calls it Buxa, *Wadia* (1957) again Baxa, *Krishnan* (1960) Buxa and *Nautiyal* (1964) Baxa. We follow the Stratigraphical Lexicon covering India, Pakistan, Nepal and Bhutan (1957) which refers with no alternative to Baxa series.)

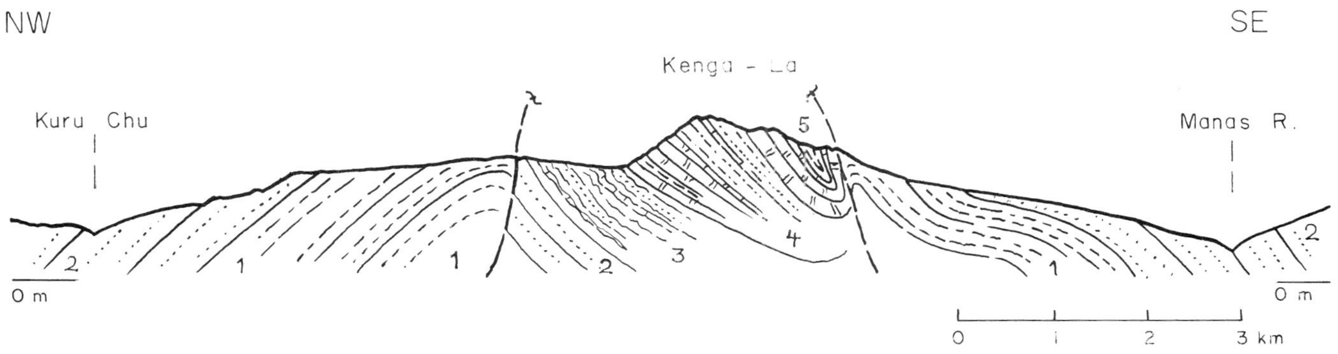

Fig.1. Section across Kenga La, between Manas River and Kuru Chu. After *Pilgrim* (1906).
1 Dalings, 2–4 Baxa quartzites, slates and dolomites.

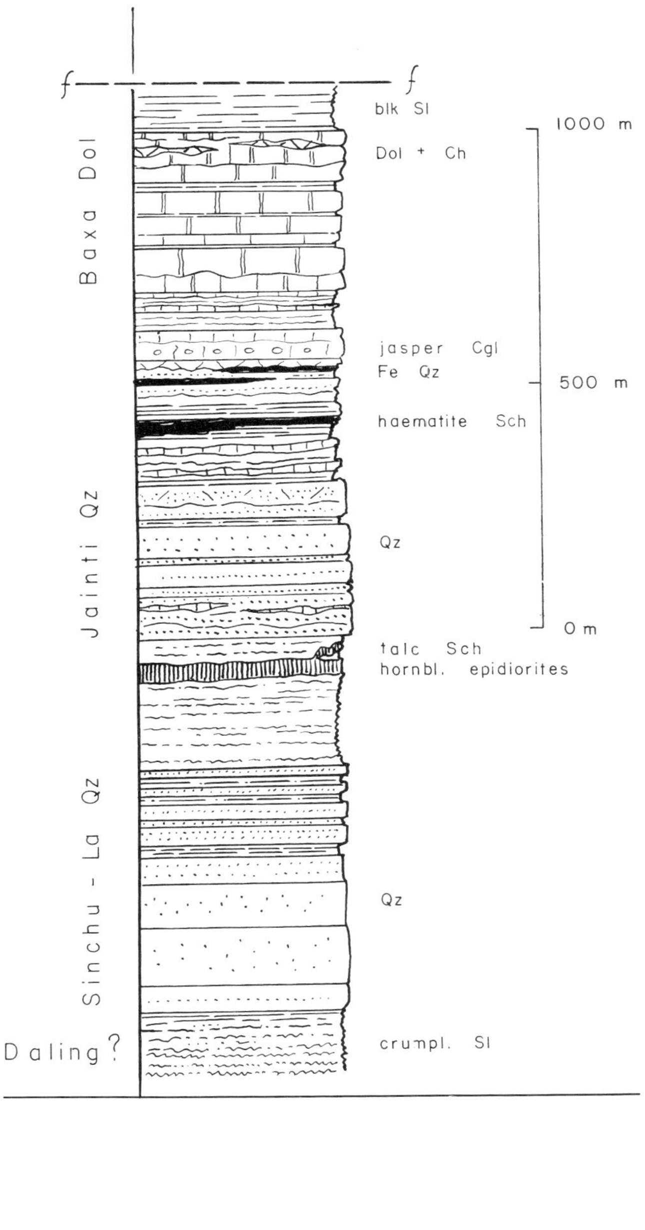

Fig.2. Stratigraphy of the Baxa formation compiled from *Mallet* (1875), *Pilgrim* (1906) and *Lahiri* (1941).

For the equivalent of the Daling, the Samchi series has been introduced. In Samchi, SW Bhutan, the outcrops are rather unfavourable for a type locality and moreover, similar to Phuntsholing, the tectonic complications are extremely pronounced along this foothill zone.

The widespread metamorphics were called «Thimphu series» while the metamorphic carbonate rocks with associated schists and quartzites are separated as «Paro series». The former locality is a rather unfavourable type section, including Paro rocks, and since the interior and particularly the northern areas of Bhutan were investigated by us, I prefer the formational subdivisions which have already been published (*Gansser*, 1964) or are proposed in the present publication.

Further geological information related to Bhutan appeared in the Abstracts of the Seminar on recent geological studies in the Himalaya (1971). K.V. *Poulose* refers to the tourmaline granite and the associated pegmatites in the Thimphus and Chekhas of Bhutan. The pegmatites are reported to be syntectonic. S.K. *Acharyya* discusses the Darjeeling frontal zone and refers to the reversed metamorphic sequence alluding to the Paro formation of Bhutan. B.S. *Jangpangi* describes two thermal springs (Rongkhola and Bhurkhola area) in S Bhutan, both related to major thrust zones.

Recently more information on southern, central and particularly southeastern Bhutan has been published by the active group of the Geological Survey of India, including data from unpublished survey reports. *Jangpangi et al.* in a paper presented 1969 but only published 1975 as well as in a more up to date paper (*Jangpangi*, 1974) discusses the lithological units introduced 1964 by *Nautiyal* et al., adding some new subdivisions. He deals with the various aspects of the Shumar formation and discusses the Diuri boulder slates which, together with his Thungsing quartzites dominate the complicated geology of SE Bhutan below the main crystalline thrust (his Thimphu formation). Of special interest are the tectonic slices of granite gneisses and granites within the Shumar formation. Still open to doubt are his age assignments, in particular the correlation of the Baxa with the Krols. A good sketch map of SE Bhutan is added.

Sengupta and *Raina* (1978) report on the Bhutan foothills adjacent to the Darjeeling district. They try a comparison of the southwestern Bhutan formations with similar and better known sections in the Darjeeling foothills. They stress the facies differences responsible for the various lithotypes within a general Daling or Baxa stratigraphic unit. This contradicts the detailed stratigraphy of *Acharyya* (1974,1975) in the Darjeeling foothills and the Baxa-Duars region of southwestern Bhutan. This author places his Baxa group above a thick Daling sequence, the latter sliced by numerous thrusts and overprinted by a syn to post genetic reversed metamorphism with isogrades cutting tectonic boundaries.

A general compilation on the «Stratigraphy and Structure of Bhutan Himalaya» by *Jangpangi* (1978) is based on all available information. He shows the Diuri boulder slates restricted to southeastern Bhutan and compares them with the Rangit pebble beds of Sikkim where fossils suggest a Permian age. From Central Bhutan he describes the Paro window and confines the highly complicated metamorphics to the southwest of Thimphu. They are suggested as a metamorphic equivalent of the Baxas. The Thimphu crystalline of *Nautiyal et al.* (1964), widespread in northern Bhutan is shown as the base of the Chekha and following Tethyan sediments (Linghshi basin). Here the survey work was only sporadic and most of the information is compiled from *Gansser* (1964). The respective structural, petrological and stratigraphical interpretations in this area by this author are still vague and not always convincing.

In addition to these strictly geological papers I refer to some geographical and morphological studies of the Bhutan Himalaya. In 1967 P.P. *Karan* published his volume on «Bhutan, a Physical and Cultural Geography». This work is based on a geographical expedition in 1964 by Indian geographers and was compiled through the University of Kentucky. I refer here to the chapter on Geographical History giving a story of the exploration and to the Physiographic Setting with a geologic description based on our 1963 expedition (*Gansser*, 1964). In a comprehensive study on the Geomorphic History and Crustal Movements of the foothills of the Himalayas by T. *Nakata* (1972), the author has included the N Bengal foothills and the Central Bhutan foothills with the Hatisar embayment, and gives a detailed account on the Quaternary geomorphology. He recognizes a multi-cyclic development of geomorphic surfaces in the Sub-Himalaya contrasting with a monocyclic surface in the plains, indicating recent upheaval in the former and subsidence in the latter. He states that the boundary between the area is clearly marked by the Himalaya Front Tectonic Line, both sides of which, the Himalayan ranges and the basement of the Indo-Gangetic plain, have reversed crustal movements that are distinctly reflected by the Quaternary tectonic movements throughout the Himalayan foothills.

More subjects related to Bhutan are found in the following publications: The accounts of M. *Ward* (1966) and J. *Tyson* (1978) deal with the northern area of Bhutan. Of topographic interest is the sketch map by Dr. *Ward* and travel companions showing the various approaches to the little known Lunana region, where they spent some winter months with high altitude medical research. A more general information is given in a book by G.N. *Mehra* written on request by the late King of Bhutan (1974). It covers the interesting period of the slow modernisation process of Bhutan. I further refer to some short notes by R. *Hänny* (1966) and B.C. *Olschak* (1971). My own previous publications on Bhutan deal with our first geological exploration (1966) and researche work on the glaciers and related lakes in the Lunana region, shown on a special sketch map (1970). The subject on these dangerous glacial lakes was further discussed in a report to the Bhutanese Government (1978). In 1971 appeared a popular book on Bhutan, coauthored by my daughter Ursula and the Tibetologist Blanche Olschak, with 80 full paged colour plates which include the still little known higher mountain area (*Gansser et al.*, 1971).

In all these publications one notes the great confusion which still exists when spelling local names. Most difficult is the proper naming of the little known mountains, since the local population is interested mostly in the passes and the mountain terminology changes pending the side you are looking at except they are the abodes of some deity. As much as possibly I have followed the advice by Blanche C. *Olschak*. Of special interest, also in this connection, is her scientific book: Ancient Bhutan, a study of early Buddhism in the Himalayas (1979). We must realize that Bhutan is the last place with an active lamaist religion.

«As a country of Himalayan mountains and highland jungles, situated between Tibet and India, near to the place of origin of the Tibetan civilisation and a living witness of its greatness, Bhutan deserves to be better known» (from the preface by R.A. *Stein*).

A.2. Present work

Our own research work in the Bhutan Himalaya began in spring 1963, while the Indian Geological Survey started field work in the foothills 1961. Up to this time, the majority of the independent Kingdom of Bhutan was geologically practically unknown, in contrast to Nepal, Sikkim and the western Himalaya where much regional and detailed work has been done. My own interest in the geology of the Bhutan Himalaya arose in connection with a general compilation on the geology of the whole Himalaya which was published in 1964, including the recent and preliminary results on Bhutan. Introduced by *Fritz von Schulthess* to the Royal Family of Bhutan, I was able to cover a large part of this fascinating country with permission and assistance of the Government of Bhutan. All available scientific and practical results of our traverses were handed to the Government.

Field geology in Bhutan is not easy. In spite of valuable assistance by the government, particularly regarding transport, food supplies and local permission, climatic conditions can handicap geological investigations considerably. Bad visibility and heavy rains in the monsoon season leave only the pre and post monsoon period for efficient work. But even in those rather short spells in the high mountains, where much of our work was carried out, sudden snow falls in the fall season can block important passes and the heavy late winter snow, particularly on the northern slopes, together with often unstable weather in the pre monsoon are equally unfavourable.

During my first and second expedition from March to June 1963 and October to December 1965, I was accompanied by my young assistent *Ruedi Hänny* who mostly undertook the structural, petrological and isotope investigations. This also included work on my samples collected on the two following trips: those I carried out from October to December 1967 and October to December 1969. After the second trip, during which the Kalanodi coal mine was investigated in SE Bhutan, it became evident that in the High Himalaya, forming the border range towards Tibet, existed various dangerous glacier lakes some of which had already caused devastating floods. This important discovery, not unlike the glacier lakes of the Peruvian Andes producing the feared «Huaicos», was included as one of the major study objects during my 1967 expedition into the wider Lunana area. The same year I handed a sketchmap indicating the most dangerous lakes with the related threatened valleys to his Majesty the King of Bhutan (see also *Gansser*, 1970). I continued these observations during my fourth expedition in the fall of 1969 into NE Bhutan, in addition to the regional geological investigations in this completely unknown area.

Newly imposed general travel restrictions (inner line control by India!) in northern and eastern Bhutan prevented further research except for my last visit so far in the fall of 1977. It is also for this reason that I decided to publish the results obtained so far, well aware about the great amount of work still needed in order to get a more complete and comprehensive geological picture of this fascinating section of the Himalayan mountains.

During all these trips possible economic deposits were studied. Many local occurrences were noted, but not unlike the Alps, larger mineral concentrations which would justify its exploitation have not been observed so far. I like to stress this point, contrary to some more optimistic reports in order to allow a realistic approach for future development (*Künsel*, 10.12.1978). Roofing slates of excellent quality could be of general interest if properly exploited and locally prepared. A special report on this item has been sent to Bhutan in early 1972. The first results obtained during the present exploitation are favourable and reserves are substantial. The larger gypsum occurrences in SE Bhutan may be of some interest as long as they can be exported to India. In a similar position are the local coal horizons along the southeastern foothill zone. Briquetting is necessary to make this highly fractured coal transportable. There also exists the potential of excellent hydroelectric possibilities, but at the present stage of development of the country larger schemes are hardly of interest for Bhutan alone.

The lack of topographical bases is a particular handicap for geological mapping. More than half of the country has now been covered by aerial photographs, but most unfortunately, they are so far not available. This also refers to the new topographical maps (1:50000) prepared from the above mentioned photographs. The old quarter inch and locally one inch maps have been used for a preliminary base but this only for the southern regions while in the north these maps are utterly wrong (S drainage cutting watersheds, see 250000 map of *Karan*, 1965 and 250000 *Bhutan* map by Survey of India 1972).

In order to locate the geological observations all traverses were measured mostly by timing distances and measuring directions with a hand-compass. These traverses were further tied in by a rough triangulation to marked morphological points. Oriented handsketches and photopanoramas improved the topographical picture. The plotting of these traverses was made in 1:100000. In 1973 the ERTS satellite photos were received covering the greater part of Bhutan and the surrounding areas. These pictures were of great help, and enabled us to adjust our local surveys into an undistorted topographical frame. They also allow the interpretation of more regional geological features. The photographs with an original scale of 1:1000000 were carefully enlarged to 1:250000 the scale selected for our geological reconnaissance map. As a reference base for the co-ordinates of the 250000 scale map the grid as shown on the 1:1000000 and 1:500000 world aeronautical chart covering Bhutan were adjusted to the center co-ordinates of the available ERTS photos. This procedure allows the incorporation of this map into the regional compilation sheets.

The political border is not indicated on our map since in various areas the frontier line is not yet properly defined in the field. The Tibetan border in particular is highly inaccurate. This is evident on *Karan*'s map where the highest mountain of this region, Kula Kangri (better Künla Kangri) is shown within Bhutan while it is actually situated completely within Tibetan territory, as already indicated on a sketchmap by the botanist S. *Nakao* in 1958.

The present account of the Geology of Bhutan is evidently a classical field report. Text and numerous fieldsketches as well as photographs are giving the facts gathered during my various expeditions into Bhutan. I am fully aware that at the present stage laboratory work, except for some chemical analysis of the leucogranites, has not yet been carried out. This refers to special petrological as well as paleontological investigations, where much more work is needed. All sketches and photographs were made by the author, except where otherwise mentioned.

I would furthermore like to stress the fact that my exploration work in Bhutan has been done entirely on my own initiative and not on the request of the Government of Bhutan.

A.3. Acknowledgments

Our geological work would not have been possible without the valuable assistance of various institutions. Our first expedition in 1963 was possible through the kind introduction by *Fritz von Schulthess* to the Royal Family of Bhutan. It was financed by the *Swiss National Science Foundation* which also has contributed substancially to the printing costs of this volume. On subsequent trips the *Swiss Foundation of Alpine Research* was assisting our work. For additional financial and scientific help which included photography, preparation of thin-sections, various laboratory investigations, draughting work and typing, the *Geological and Petrological Institutes of the ETH and University of Zürich* and its staff are specially mentioned. *David Ellis* and *David Rubei* have kindly read and corrected the text. The great help of all these institutions is highly appreciated. I thank the *Denkschriften Kommission* of the S.N.G. who has accepted this work in her series of publications and the Editors of the *Birkhäuser Verlag* for the efficient cooperation.

I greatly regretted that my companion on the first two expeditions to Bhutan, *Ruedi Hänny* was prevented through other assignements to complete the present publication. Much of the petrological and structural information given herwith is based on his research.

The greatest assistance, however, has been received by the *Government and the People of Bhutan*. To the *Royal Family* with *His Royal Highness Wangial Wangchuck* and particularly to *His late majesty King Jigme Dorje Wangchuck* we owe our special thanks. The untimely death of an unforgettable humane ruler is greatly regretted.

To his memory the present work is dedicated.

B. The Geological Subdivisions of the Bhutan Himalaya

A geological subdivision is based on various criteria. Stratigraphical as well as structural aspects are involved. Both are in Bhutan still not sufficiently known and our proposal is only a preliminary one and open to further refinement and evidently also changes. The Geological Survey of India (*Nautiyal et al.*, 1964) as well as *Gansser* (1964) have published nearly at the same time a subdivision into geological units. The one proposed by us is in a way less detailed than the one by the Survey geologists. On the other hand not all of these proposed subdivisions are based on well established type sections and are stratigraphically still disputed. Since both are meant for a working hypothesis and a preliminary base for further work, the main divisions proposed here are based as far as possible on both publications.

For the whole Himalayan chain I have proposed from S to N the following main zones (*Gansser*, 1964) (fig. 3):

1) A Sub-Himalaya *with a belt of molasse like clastic deposits, the Siwaliks, which border the Foreland basins.*

2) The Lower Himalaya *with huge sedimentary sections of mostly late Precambrian age coverd by Gondwana type rocks and by crystalline thrust sheets exposing a reversed metamorphism.*

3) The High Himalaya *consisting of thick crystalline thrust sheets which form the base of the Tethyan sedimets to the north.*

4) The Tibetan Himalaya or Tethys Himalaya *with an independent tectonic on top of the crystalline sheets and involving a conformable stratigraphical column from late Precambrian to Eocene, and lastly,*

5) The Indus-Tsangpo zone, *a major suture zone displaying orogenic sediments, ophiolites and ophiolitic melange formations with exotic blocks and large ultramafic thrust sheets.*

In the Bhutan Himalaya only four of these five zones are more or less well defined. The Indus zone follows far to the N along the Tsangpo river within southern Tibet and was out of reach for our investigations except for a visit during 1980 (*Academia Sinica,* 1980) (Fig. 136, 137). Sediments representing the Tethys or Tibetan Himalaya occur only as basin-like embayments in northwest and central Bhutan and as a thin margin just along the Tibetan border in north-central and northeast Bhutan. The larger part of Bhutan consists of crystalline sheets representing the crystalline of the High Himalaya including, towards the N, larger masses of Tertiary granite intrusions. Difficult to place is the southern limit with the fronts of the crystalline thrusts morphologically already within a Lower Himalaya.

We note similar conditions in eastern Nepal, Sikkim and Arunachal Pradesh (NEFA) where the crystalline belts encroach upon a narrow sedimentary Lower Himalaya. The latter form a highly complicated but locally well developed belt in southern Bhutan, including the mostly carbonaceous sediments of the classical Baxa hills, higher sections of quartzites and slates and thin wedges of highly tectonized coal bearing zones recalling the Permo-Carboniferous Damudas. With a rather steep contact they are thrust over the molasse rocks of the Siwaliks (Main Boundary Thrust). These form a surprisingly constant belt all along the 2400 km Himalayan foothills, except for the Bhutanese part where some conspicuous gaps do exist, while further to the E, along the Arunachal Pradesh (NEFA) foothills, the Siwalik beds are again well and continuously exposed. Large post-Siwaliks, Pleistocene to Recent detrital fans cover or partly replace the Siwaliks along the N edge of the Assam plain. Locally they are tilted and in the Jaldhaka river area at the southwestern border of Bhutan they are clearly warped into broad anticlines (*Heim* und *Gansser*, 1939) (fig. 4).

In the following I will first discuss the Sub-Himalaya, then the Lower Himalaya and the Tethys Himalaya before dealing with the crystalline nappes (High Himalaya) which cover the greater part of Bhutan. The term Higher Himalaya is here less adequate since morphologically part of the crystalline nappes cover also the Lower Himalaya. This chapter is followed by discussions on the metamorphic grades, the structural aspects, glaciology and the final conclusions.

Fig. 3. Geotectonics of the Wider Himalayan Region

Fig. 4. Warped terraces in the Jaldhaka River area after *Heim* and *Gansser* (1939).

◀ to Fig. 3.

Abbreviations

Bo	Bombay	Ls	Lhasa
Ca	Calcutta	Mu	Muscat
De	Delhi	NB	Namche Barwa Mt.
He	Herat	NP	Nanga Parbat Mt.
Is	Islamabad	Qt	Quetta
Jl	Jolmo Lungma (Everest)	Ra	Rangun
		Sp	Spongtang Klippe
Ju	Jungbwa nappe		
Kb	Kabul	*main structural units*	
Ks	Kailas Mt.	SZ	Suture Zone (Indus/Tsangpo)
Ka	Karachi		
Kt	Kathmandu	MCT	Main Central Thrust
Ko	Khotan	MBT	Main Boundary Thrust
La	Ladakh		
		MFT	Main Frontal Thrust

Outside wider Himalaya and Indian shield only structural trends, ophiolites and volcanoes are shown.

- Al — Alluvial plain
- Pre-Gondwana basement, Indian shield
- Gondwana sediments on Indian shield
- Mesozoic Platform sediments on Indian shield
- Trap, Indian shield
- Molasse type sediments (Siwaliks of sub-Himalaya)
- Lesser Himalaya
- Crystalline of High Himalaya
- Tethyan sediments (platform)
- Flysch facies
- Ophiolites incl. ophiolitic melanges and related pelagic sediments
- Transh malayan plutons
- Tertiary and Quaternary volcanics
- Quaternary volcanoes
- Main thrust and fault zones
- Lineaments (fold axis, sec. fault zones, fracture zones)
- outline of Bhutan

C. The Sub-Himalaya of Bhutan

The Sub-Himalaya consists here of the Siwalik molasse and younger Quaternary fans and terraces. The northern border is marked by the mostly steep Main Boundary Thrust and the southern limit by encroaching alluvial deposits. The latter can reach the Main Boundary Thrust in those sections where the Siwaliks are missing. This latter fact is most conspicuous for parts of the Bhutanese foothills and contrasts with the otherwise practically constant Siwalik belt all along the Himalayan chain.

C. 1. The Siwaliks

From the Tista river in Sikkim, where over 3000 m of well developed Middle to Upper Siwaliks outcrop, one can follow these horizons eastwards over a small gap at the Chel river to the Jaldhaka river. They strike here with a ENE direction directly into the promontory of Baxa and older formations of the Samchi hills of southwestern Bhutan. This southwards bulging thrust zone of the Lower Himalaya seems to have overridden an erosional gap in the Siwaliks producing a so-called relief-thrust, well-known in the Subalpine molasse belt of the Alps and which was suggested already for the region of Nainital (*Heim*, 1938; *Gansser*, 1964) (see Geological Map Plate I which is enclosed in this book).

Further to the E, Siwaliks beds occur again E of Phuntsholing along the Baxa Duars. They crop out in the lower Wang Chu or Raidak river with beds dipping 80° to the N. A few hundred meters to the south of this Siwalik outcrop *Mallet* (1874) made an important discovery of an outcrop of basement rocks forming a «normal» base to the Siwaliks. On page 44 he states: «Proceeding up the left bank of the Raidak river, the first rock met with is hornblende schist in thick beds dipping north 10° west at 50°, and forming a low eminence; only a small thickness is seen. Beyond this is blank for a couple of hundred yards, and then Tertiaries come in, dipping locally to north at 80°. There can hardly be a doubt that this rock belongs to the gneiss which forms most of the hills that are scattered over the alluvial valley of the Brahmaputra, and which, according to Mr. Medlicott, there is no reason to suppose its distinct from that of Bengal. The southerly trend of the Tertiary hills both east and west from the debouchure of the Raidak make it probable that the Tertiaries are not faulted against, but overlie the gneissose rocks. The above outcrop is the one solitary point at which the Assam gneiss has been observed at the foot of the hills between Nepal and the Goalpara district.»

This discovery, if further verified, would be of far reaching consequence, as it is the only spot along the whole Himalayan range where basement rocks belonging to Peninsular India (here Shillong plateau) crop out immediately adjacent to the Siwalik belt. It would be a further proof that the Shillong basement mass reaches practically the Himalayan foothills and that no foredeep is present along this part of the mountain front. Subsequent investigations by geologists of the Indian Geological Survey (*Acharyya*, 1971) cast some doubt on *Mallet*'s assuption of the existence of Peninsular shieldrocks. From the new description it is however not quite certain if the same locality has been visited, since *Mallet's* outcrop was reported several 100 m below (south) of the Siwalik front. On the other hand river erosion can considerably change the outcrop pattern, especially when considering a span of about 140 years. Reported are 3 slabs of plagioclase amphibolites (4 × 1,5 m) interbedded in pebbly Siwalik sandstone, not mentioned by *Mallet*. The flattish amphibolite slabs are supposed to have «skidded» from intrusions within the Baxa Group about 8 km to the N. *Lahiri* (1941) does record epidiorite sills and dykes from the Sinchu-La stage but not from the Baxa. He also collected specimens of some gneisses of Darjeeling type from the Raidak gorge. Though a transport from the N into the present position seems likely, the fact that these crystalline rocks occur in an area where Shillong shield rocks outcrop less than 50 km from the Himalayan front is still of some interest.

From the Manas River to the E the Siwaliks reappear as a constant belt of variable width for over 500 km to the Lohit area (Mishmi hills) in easternmost Assam. The best Siwalik section in Bhutan is outcropping east of the Manas river in the Samdruk Jongkhar-Dewathang area (fig. 5). In the Kalanodi coal area in the most southeastern part of Bhutan they are again tectonically reduced but further E they increase forming a substantial belt at the Bhareli river along the Arunachal Pradesh (NEFA) foothills.

The various Siwalik outcrops along the Bhutanese foothills have a more or less steep northerly dip and include the well exposed Dewathang Section, which seem to mainly belong to the Middle and Upper Siwaliks. Since deltaic cross lamination clearly indicates a normal position, the youngest outcrops occur near the Main Boundary Thrust and the oldest along the southern margin (fig. 5). According to *Mallet* (1875) and *Pilgrim* (1906) who respectively described Siwalik sections in western and eastern Bhutan the 3000 m deposits represent the Middle and Upper Siwaliks. On the other hand *Nautiyal et al.* (1964) recognize the classical three fold division, but they mention however, that the distinction between

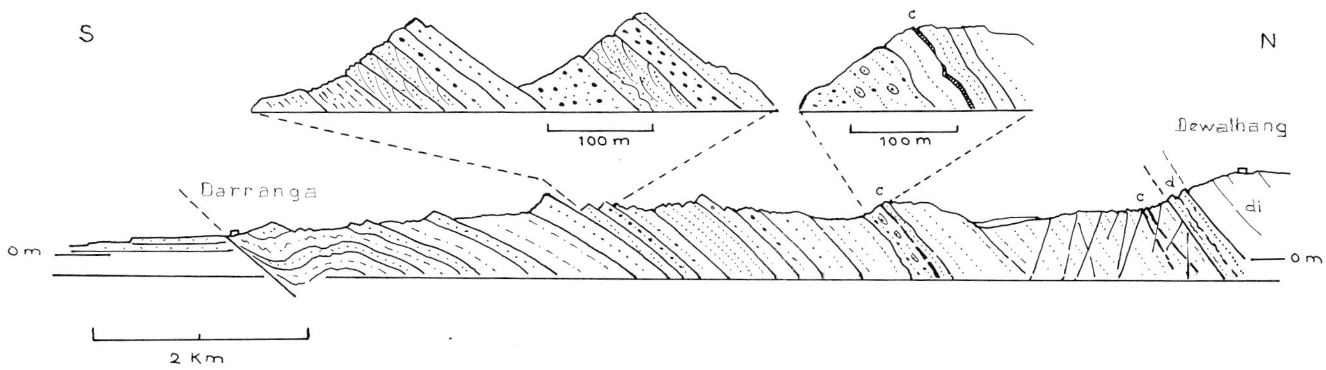

Fig. 5. The Siwalik section S of Dewathang with inset detailed profiles. (SE Bhutan) *c* lignitic coal, *d* Damudas, *di* Diuri formation.

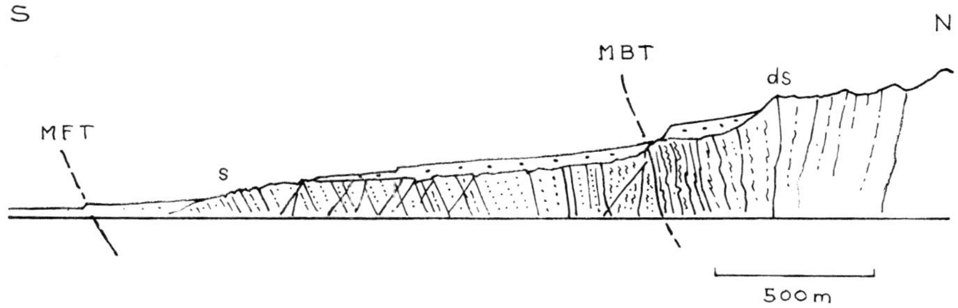

Fig. 6. The Siwalik section along the lower Manas river. Shows strong cleavage and small faults.
s Siwaliks covered by tilted terraces, *ds* Daling-Shumar group.

the Lower and Middle subdivisions is not clear. *Godwin-Austen* was the first to visit the Siwaliks of the Baxa Duars (1868). He even produced an elephant molar tooth and observed further the widespread well developed younger terraces. He also noted the disappearance of the Siwaliks some kilometers W of Baxa. *Pilgrim* (1906) studied the Siwaliks in 3 sections east of the Manas river, including the one south of Dewathang visited by us. He stresses the great regularity of the northerly dips with 50° to 60°, and records clays and finegrained sandstone in the southern (older) section grading over micaceous «salt and pepper» sandstone into conglomeratic sandstone, the pebbles of which consist invariably of quartzite while gneisses are missing. Stringers and nests of lignite were observed in these higher horizons.

At the Manas river a small section of Siwaliks is exposed below several stages of river terraces (fig. 6). The Siwaliks consist predominantly of siltstones alternating with argillaceous sandstones dipping steeply to the N. The beds seem overturned locally towards the Main Boundary Thrust, which is however badly exposed. Near the contact, highly tectonized and sheared sandstones contain small pebbly stringers. A marked cleavage and small southwards trending faults cut the steep sediments which may belong to the Middle Siwaliks.

Along the Dewathang section (fig.5) the exposed Siwaliks are about 4000 m thick, assuming the deposits to be normal. The Main Boundary Thrust is not well exposed. The siliceous slates of Dewathang belonging to the Lower Himalayan sequence are phyllonitized and border against whitish quartzitic sandstone which *Nautiyal et al.* (1964) separate as «Dewathang Series» and compare them with Lower Tertiary Subathus. This seems doubtful since these sandstones are followed by black mylonized carbonaceous shales, dipping with 70° to the N, the trend of the Main Boundary Thrust. This section is more reminiscent of the Damudas, thrust steeply against the Siwaliks. These begin with green sandstones, silts and thin slickensided coal seams and coaly fragments. A similar section with coaly fragments and badly preserved plant remains within the Siwaliks is exposed near the Main Boundary Thrust in the Kalanodi river in the southeastern part of Bhutan. The northern Dewathang section is generally highly fractured and may contain tectonic complication including some older Siwalik slices. Only 1½ km southward begins the normal sequence with micaceous sandstones, some large sandstone concretions and irregularly disseminated pebbles, consisting mostly of quartzites. Real conglomerate horizons occur only in the middle part of the section but here with smaller pebbles, rarely surpassing 5 cm. Silty layers and thinly bedded sandstones increase southwards. Fine washouts and a clear crossbedding indicate a normal section. Southwards and towards the lower part of the section bluish gray siltstones and claystones increase, alternating with fine-grained and wellbedded sandstones suggesting the Middle Siwaliks. At Samdruk-Jongkar (Daranga) they form the deepest outcrops in a small fold and then rise again southwards «against» the alluvial terraces. Our observations confirm the investigations by *Pilgrim*, mentioned above.

Jangpangi (1974) follows *Nautiyal* suggesting a

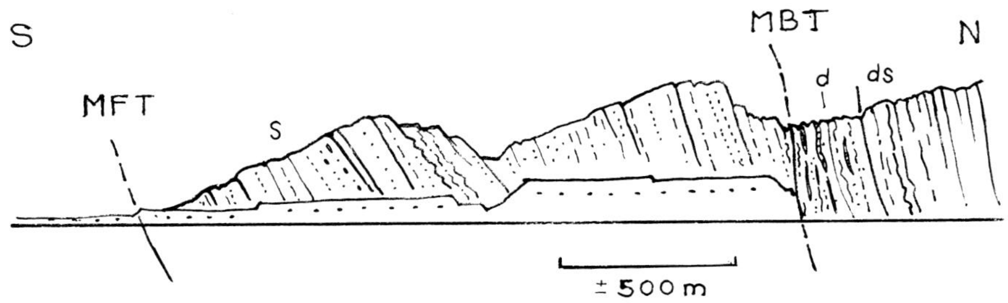

Fig. 7. The Siwaliks along the Kalanodi river (SE Bhutan) (in this section the Diklai boulder beds are not exposed).
s Siwaliks, *d* Damudas, *ds* Daling-Shumar group.

three fold division for southeastern Bhutan, in addition to the locally developed Diklai boudler bed. He also reports some wedges of «Lower» Siwaliks along the Main Boundary Thrust in the easternmost sections. The Diklay boulder bed seems a younger fan probably transgressing the uppermost Siwaliks. It is distinguished by its more fanglomeratic aspect and the yellowreddish colour contrasting with the monotonous gray of the Upper Siwaliks. It shows a weaker consolidation and the matrix is a coarse sand. It is developed E of the Dewathang section.

Further E, Siwaliks again outcrop along the Kalanodi river, well-known for the local coal mines. Here sandstones predominate, partly with small pebble layers, some larger concretions and with intercalations of bluish gray siltstones (fig. 7). They resemble somewhat the higher part of the Dewathang section. Towards the Main Boundary Thrust occurs a tectonized wedge of fine micaceous sandstones with some larger deltaic features and conspicuous horizons of claystone pellets. They are bordering fine sheared lignitic coal seams. The contact against the highly sheared coalbearing quartzites of the Damudas is sharp and about vertical (M.B.T.). Between this tectonized Siwalik wedge and the more normal Siwaliks further to the S we observed a thick formation of badly sorted, loosely cemented, coarse fanglomerate which dips gently southwards and transgresses the much steeper Siwaliks. We may have here the equivalent of the Diklai boulder beds of *Jangpangi* (1974). Of special interest is a sharp, well exposed tectonic contact between the Siwalik wedge and the fanglomerate. Siltstones and fine sandstones with thin and sheared coaly seams are faulted against the fanglomerate. Along this faultzone, strong movement has even sheared a quartzite boulder of head size completely in half, which is firmly embedded in the fanglomerates (phot. 1,2). We witness here a very marked post Siwalik movement of Quaternary age (see below) (fig. 8).

From the description of *Mallet*, *Pilgrim*, *Nautiyal*, *Jangpangi*, and our own observations, the classical three fold division of the Siwaliks is not easily recognizable. The predominance of bluish gray clays and silts over the reddish brown clays of the more western sections may indicate some different environmental conditions such as climatic changes not unlike the present greater humidity and possible stronger erosion in the eastern Himalaya. The exposed Siwalik sections along the foothills in Bhutan are mostly normal, without tectonic complications except for some local faulting and secondary thrusting near the Main Boundary Thrust. This differs from the better exposed Siwalik sections along the southeastern foothills of Nepal, where within the much larger belt steep thrustzones repeat the sections, locally exposing older Siwaliks along the Main Boundary Thrust (*Itihara et al.*, 1972).

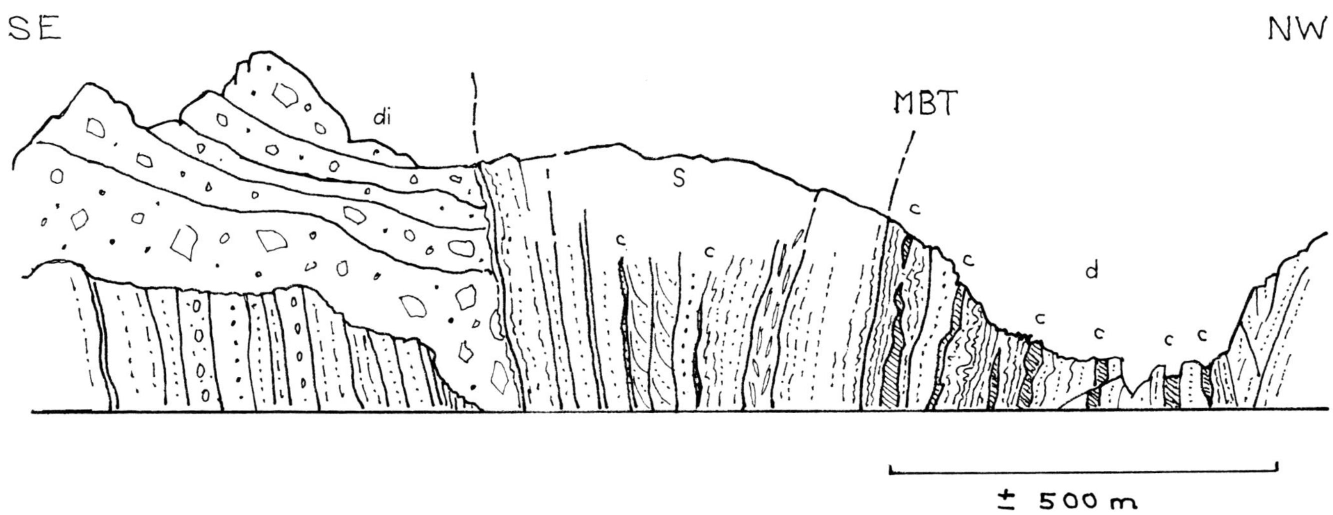

Fig. 8. Diklai boulder beds (*di*) transgressing Siwaliks (*s*) and faulted against Siwaliks along possible branch of MFT. MBT separates Siwaliks from Damudas (*d*) (*c* are coal beds).

Phot. 1 Siwaliks (1) thrust on Diklai boulder bed (3). Along thrust sheared lignitic shales and calcarous sandstones (2). Note head sized quartzite boulder sheared in half along thrust. Kalanodi river.

Phot. 2. Crossbedded micaceous sandstone of Siwaliks, fractured and sheared near MBF. Kalanodi river.

C.2. Post-Siwaliks

Apart from the Siwaliks the Sub-Himalaya is noteworthy for the abundance of subrecent to recent river terraces which were clearly affected by the youngest tectonic displacements. The fanglomerates from the Kalanodi river have shown Quaternary faulting. Large river terraces west of the Bhutan foothills (Jaldhaka river) are warped to form gentle, practically recent anticlines (fig. 4). Steplike terraces along the main rivers within and at the south margin of the foothills reflect the youngest uplifts of the Morphogenic Phase. It is further striking that the material of recent river fans and terraces is considerably coarser and badly sorted to fanglomeratic if compared to the youngest deposits of the Siwaliks. A much stronger relief, erosion and transport has to be assumed during the Post Siwaliks to Recent period and all morphological indications suggest that the related vertical movement is still most active.

It is interesting to note that the best visible Quaternary tectonics occur in certain gaps where the Siwalik belt has been interrupted by previous erosion, relief-thrusting or drowning. One such gap, at the western side of the Bhutanese foothills is well known for conspicuous Quaternary tectonics. Here tilted terraces and anticlinal upwarpings are seen along the Jaldhaka river and have already been reported by *Heim* and *Gansser* (1939) (fig. 4). This area was studied in more detail by *Nakata* (1972). Within the erosional gaps where the Siwalik sediments are missing, *Nakata* has observerd the Daling type phyllites thrust on top of the gravel terraces (Neora and Chel rivers). Based on his detailed terrace analysis he recognized renewed movements along this important thrust, reflecting several Quaternary phases. This tectonized belt is limited to the S by *Nakata*'s Himalayan Front Tectonic Line (my M.F.T.), visible along the sharp Chalsa cliff. Similar Quaternary events were observed furhter eastwards in south central Bhutan.

A pronounced gap in the Siwalik belt occurs in the Hatisar (Geylegphu) embayment, between the Sankosh and the Manas rivers. The Siwaliks as well as the Main Boundary Thrust and part of the Lower Himalayan formation are drowned and covered by alluvials in form of flat fans. It should be emphasized that no large river system is at present entering the Hatisar embayment. This area along the Bhutanese border has also been investigated by *Nakata* (1972) during his regional studies on the Geomorphic History and Crustal Movements of the foothills of the Himalaya. In the western part of this embayment and S of the assumed Main Boundary Thrust *Nakata* reported a W to E aligned row of southwards tilted gravel terraces with a sharp and steep north scarp 6–12 m high (fig. 9). The top of the terraces is covered by a red weathered soil. They are cut by antecedent local rivers and on their south-side even the youngest terraces are tilted, indicating recent movement. The sharp E–W aligned northern scarp coincides with a structural line bordering the submerged Siwaliks to the south. Conformable with the general outline of the Himalayan foothills, the upheavel and tilting of the Quaternary terraces lies here to the south of *Nakata*'s so-called Himalayan Front Tectonic Line (my Main Frontal Thrust). This example seems to indicate that the youngest upheaval was preceeded by a subsidence responsible for part of the Hatisar embayment. It is of special interest to note that this area lies just N of the above mentioned widespread basement hillocks, rising out of the Assam plain north of the Brahmaputra river and which outline a northwestwards trending spur of the Shillong shield.

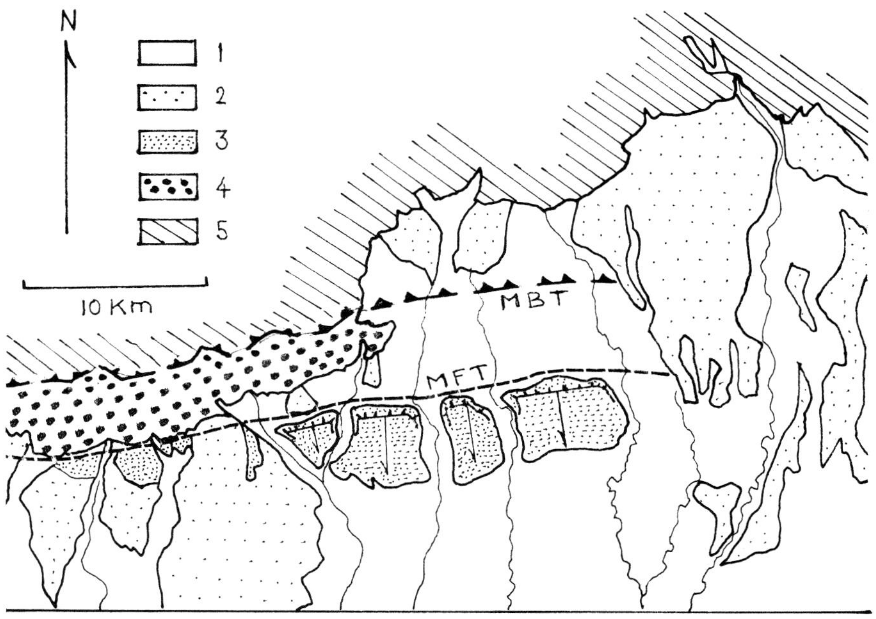

Fig. 9. Quaternary tectonics at the Hatisar (Geylegphu) embayment.
1 Alluvial fill, *2* main terraces, *3* tilted and faulted terraces, *4* Siwaliks, *5* Pre-Siwaliks, *MBT* Main Boundary Thrust, *MFT* Main Frontal Thrust (Based on Nakata 1972).

D. The Lower Bhutan Himalaya

This part of the range is the geologically most complex. Its southern border is very clearly defined by the Main Boundary Thrust but its northern limit is vage and difficult to trace. The Main boundary Thrust is exposed in some river sections and along the surface one notes a steep northwards dipping, partly even vertical thrustzone. From all evidence this thrust seems to flatten with depth. While windows of Damudas occur below the next inner thrust zone, no Siwalik windows are known so far all along the Himalaya. Relief-thrusts are present however in various areas and show thrusting of Lower Himalayan units over partly eroded Siwaliks. Of regional importance is the fact that usually the youngest part of the Siwalik section is bordering the Main Boundary Thrust, and that the Siwalik sections are mostly normal and not overturned. Quite different are the conditions north of the Main Boundary Thrust. Various units, normally divided by further, more interior thrusts can border and actually take over the Main Boundary Thrust. This is particularly true for the Permo-Carboniferous Damudas (Gondwanas) which form an intermittent belt along the Bhutanese Lower Himalaya and which are often completely cut out by the next inner thrust, bringing the Precambrian sediments such as Dalings in the widest sense as well as the only locally developed Baxas against the Siwaliks (see Geol. Map Plate I).

D.1. The Damudas, Gondwanas (Permo-Carboniferous)

The Damudas, representing the Gondwanas, are the youngest formation within the Bhutanese Lower Himalaya and occur as tectonically very disturbed, thin, intermittent wedges just north of the Main Boundary Thrust. They strongly resemble the Damudas of the Darjeeling area where they are known since *Mallet*'s descriptions of 1875. Along the Bhutanese foothill zone they have been reported from the Samchi region (*Nautiyal et al,*, 1964) from the Baxa hills (*Lahiri*, 1941) from eastern Bhutan by *Pilgrim* (1906) and by *Jangpangi* (1974). Owing to the extreme tectonization and imbrication, normal sections are rare.

The best outcrops are reported by *Jangpangi* (1974) along rivers to the E of Dewathang. From the Main Boundary Thrust to the N he mentions (p.128) «quartzitic sandstones with coal beds and carbonaceous slates, followed by calcareous slates, marly limestones, dark gray carbonaceous slates with boulders and tectonic leses of quartzitic sandstones (Gondwana) and dolomite (? Baxa) and finally dull greenish splintery shale, gradually merging into boulder slates. The coal bearing quartzitic sandstones are the most constant element in the various sections».

In the Kalanodi area we noted gritty, micaceous and feldspathic quartzites with highly sheared carbonaceous shales and irregular bands and lenses of an anthracitic coal. A strong boudinage of the quartzite layers is often visible where the carbonaceous shales dominate. The northern and southern contacts are purely tectonic. In the Kalanodi river irregular coal pockets are mined. The average coal sample shows a fixed carbon content of about 50% with volatiles averaging 25–30%, an ash content of about 25% and 5700 to 7000 cal. The highly sheared coal can only be extracted in small fragments which further disintegrate during transport. Briquetting the coal seems the only solution and tests of the Kalanodi coal gave favourable results in this respect.

The thin and highly tectonized Damuda outcrops are easily eroded and often covered by scree so that they may actually be more continuous than what can be deduced from the scattered outcrops. Along the thrustzone on the north side of the Damuda belt, Baxa and/or Daling formations are thrust over or against the steep Damudas. It seems likely that the hade of this thrust averaging 50° N along the surface may flatten at depth. This possibility could be assumed for the foothill zone of Sikkim where the Damudas reappear 30 km inland in the Rangit window (*Gosh*, 1952) (see Plate I).

D.2. The Diuri Formation

We deal here with a peculiar section from SE Bhutan which is well exposed to the N of Dewathang along the Dewathang-Tashigang road, and forms a broad, partly faulted synform. The over 2000 m thick section is characterized by pebbly shales and slates and certain outcrops are not unlike the Blainis from the western Himalaya (Mussoori-Rikikesh area). *Nautiyal et al.* (1964) refers to it as Diuri Boulderbed and stress the similarity with western Himalayan Blainis and places them into their Baxa Series. This last suggestion is difficult to follow. I accept *Nautiyal*'s term Diuri but prefer to regard it as a separate unit: the Diuri Formation between the Damudas and the more locally developed Baxas and/or older deposits.

In eastern Bhutan the Diuri Formation begins approximately 12 km N of Dewathang and reaches southwards to the Main Boundary Thrust. In the N it transgresses with a badly sorted conglomerate on highly sheared greenish quartzites belonging to the older Daling formation (the Thungsing quartzites of *Jangpangi*, 1974). The contact seems primarily discordant but has subsequently been tectonically disturbed, so that the original relations are not clear. The pebbles vary from 0,5–5 cm and are poor to well rounded, depending on the rock type. White and dark grey quartzites dominate, less frequent are white vein quartz, some siliceous yellowish dolomite, black irregular slate fragments and, curiously enough, some small rounded pebbles of perthitic alkalifeldspar crystals with some quartz. The matrix consists of a fine sericitic quartzite. Above this conglomerate follow over 2000 m of black to dark ash grey slates and pebbly shales with silty and fine quartzitic bands. The ground mass of the pebble beds is mostly phyllitic with fine disseminated sericite, muscovite, chlorite, small quartz and calcite grains and with a more or less pronounced graphitic substance. The latter can locally be enriched in lenses, but no real coaly matter has been observed. The ground mass shows a clear schistosity and is often fluidally arranged around the lenticular pebbles (phot. 3). These consist predominantly of white quartzites, siliceous dolomites, black slates, and smaller pebbles of perthitic alkali feldspar and plagioclase quartz fragments. About 1,5 km N of Dewathang occur larger pebbles (3–8 cm ∅) of fine grained dolomite. They show a thin rim of a fibrous calcite with lamellae parallel to each other and sharply discordant to the pebble surface. Their outer end is curved and suggests a rotation of the pebbles. The optical orientation clearly indicates a recrystallisation after shearing. Many of the irregularly disseminated pebbles show a surprising «Dreikanter» form. One may think of wind erosion within a glacial climate prior to the sedimentation in shallow water.

Phot. 3. Tillitic aspect of Diuri boulder bed. Irregular quartz grains, phyllitic matrix. S. Narphong.

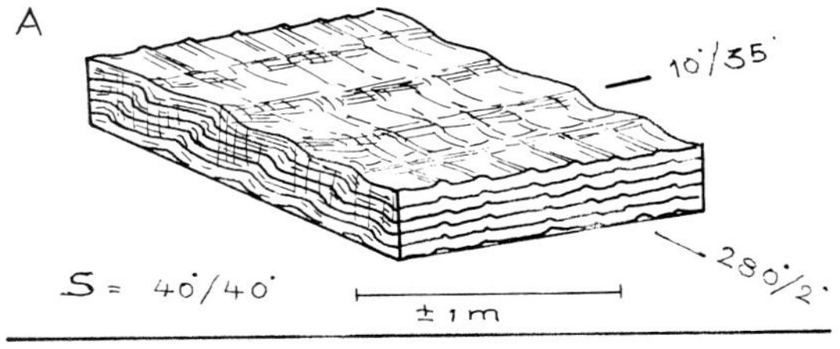

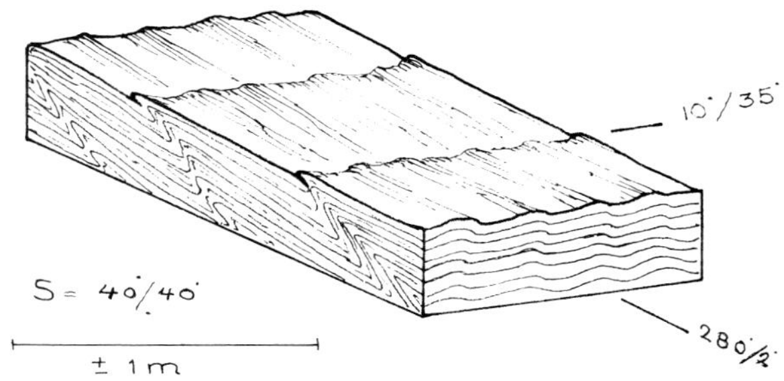

Fig. 10. Folds, lineations and overprinted kinkbands in the Diuri phyllitic slates. The overprinted structures parallel the Kuru Chu-Shumar spur.
A = weaker, *B* = stronger overprint.

The dominating slates and phyllites show a strong lineation with overprinted kinkbands (fig. 10). Apart from the «Dreikanters» the pebbles are generally stretched parallel to the main lineation which strikes E–W in the southern part and changes to NW in the more northern section of the Diuri boulder slates.

The lack of coal, the great thickness, the rare gritty quartzite intercalations as well as the incipient metamorphism suggest an older age of the Diuri formation if compared with the Damudas of the foothill zone. They are further separated by a marked thrust and belong structurally to a more internal zone. On the other hand, there is some lithological similarity with the Blaini pebble beds from the western Himalaya but long range correlations within the peculiar Lower Himalaya facies belt are most speculative. A palynological check on the more carbonaceous pebbly samples turned out negative. The possibility of Late Precambrian-Cambrian tilloide deposits should not be completely excluded (*Bhatia* and *Kanwar*, Ed., 1975).

D.3. The Baxa Formation

Carbonate rocks, so far unfossiliferous except for algal beds such as Collenias, are widespread in the Lower Himalaya and can represent sections of several 1000 m (*Fuchs* and *Sinha* 1978; *Valdiya*, 1969; *Gansser*, 1964). The facts that they occur in various tectonic units, that they are subject to rapid facies changes and, except for stromatolites, are mostly without determinable fossils, makes their correlation within the Lower Himalaya belt most difficult. A certain structural unit has been separated since the regional investigations by *Pilgrim* and *West* (1928) as Krol belt and was subject to many subsequent investigations and discussions (*Auden*, 1934, 1970). The type Krol belt is certainly younger than the carbonate formations in the Bhutan Lower Himalaya and does not extend to the eastern Himalayas at all. The correlation of the Baxa with Krol as suggested by *Nautiyal et al.* (1964) is open to doubt.

Mallet (1875) named a dolomite, shale and quartzite sequence between the Gondwanas and the Dalings Baxa Series after the old Baxa Fort in the western Duars of southwestern Bhutan. A similar section was recorded by *Pilgrim* (1906) in south central Bhutan. *Nautiyal et al.* (1964) stressed the lenticular and en echelon pattern of the various Baxa outcrops but includes the Boulder Bed of Diuri.

In the Lower Himalaya of Bhutan it is quite impossible for me to separate the larger quartzite, slate and shale sections placed into the Baxa by previous authors from similar horizons outcropping in the Phuntsholing section. I prefer, therefor, to limit the Baxa formation to those horizons which contain mostly dolomites, limestones, calcareous shales and locally gypsum and which begin with basal quartzite beds, the Jainti quartzites of *Lahiri* (1941). Baxas are known from the various scattered outcrops in the Samchi area (*Nautiyal et al.*, 1964), the Baxa Duars (*Lahiri*, 1941), the Phuntsholing section, the lenticular masses W and E of the Hatisar embayment, the Kenga La section (*Pilgrim*, 1906) and the various outcrops in southeastern Bhutan discussed and presented on a small map by *Jangpangi* (1974, 1978).

The base of the Baxa formations (or Baxa Group since it consists of various mappable units) are flaggy, irregular banded quartzites which often have sericite talc-schists at their base. The upper quartzite is more massive, with sericite and chlorite, and locally irregularly disseminated copper ores (chalcopyrite). The quartzites are overlain by banded micaceous haematite quartzites (Jtabirites) and red jasper which locally develops into jasper conglomerates with a quartzitic matrix. These rocks are of particular interest since they could be compared to the well known banded ironstones of late Precambrian age. A Permo-Carboniferous age for the Himalayan banded ironstones as suggested by O'*Rourke* (1962) is unlikely for the present section. The most conspicuous part of the Baxas, however, are the carbonate rocks overlying the ferruginous horizons. They consist of several dolomitic limestones and thick dolomite layers. The lower dolomite horizons are light-grey and thick-bedded, while the upper horizons are thinner, dark grey and often of a saccharoidal aspect. A slight mineralization by galena and sphalerite has been noted in the lower layers (*Lahiri*, 1941). The carbonates are practically free of siliceous impurities. The uppermost dark slates are cut off by the thrusts which carry the older Daling phyllites and shists over the Baxa outcrops.

Along the Phuntsholing section (fig. 11) only the part below the carbonate Baxas are exposed in the form of reddish mudstones and shales with haematitic schists and a corresponding thick ferruginous weathering. They are followed by ferruginous banded quartzites, some jasper (chert) horizons and variegated red and green shales. This colourful section is disharmonically folded with flat SE-NW directed foldaxes. Steep northwards heading thrustzones have cut out the carbonate Baxas. Within the following imbrications occur thicker quartzites which contain copper ore. They alternate with variegated shales and expose highly sheared talc schists at the base. They are similar to Jainti quartzites of the Baxa type section (fig. 2).

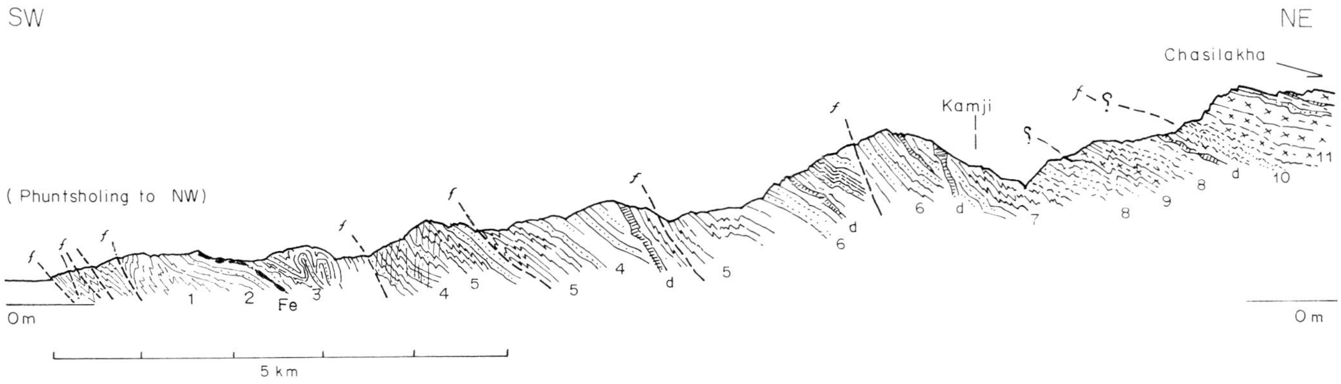

Fig. 11. Section across the SW Bhutan foothills (Phuntsholing-Chasilaka)

1 strongly sheared slaty phyllites,
2 red claystone with haematitic schists,
3 ferruginous quartzites with red a. green shales,
4 dark gray phyllonites and sheared quartzites,
5 green and violet slaty phyllites with cherty quartzites,
6 thick bedded white quartzites with sericitic and talcose partings (Cu ore),
7 folded schistose phyllites,
8 sericite schists, and quartzites
9 sheared augengneiss,
10 biotite-sericite schists,
11 biotite-granite gneiss (Chasilakha zone),
d hornblende diabases.

Larger lenses of km size of Baxa dolomites have been reported by *Jangpangi* (1974,1978) east and west of the Hatisar (Geylegphu) embayment. They may be related to the peculiar conglomerates near the hot springs of Mao Khola (Rong river), described below. Tectonized Baxa rocks outcrop along the lower Manas river in several, steeply northwestwards dipping layers, their strike suggesting already the large Kuru Chu half window or spur (*Jangpangi*, 1974).

In southeastern Bhutan a tectonic wedge of Baxa type dolomites is crossed by the Tashigang/Dewathang road just north of Narphong (km 59). The dolomites occur between normally northdipping quartzites and phyllites of the Daling/Shumar types. At the tectonic contacts the dolomites are bordered by strongly sheared, partly red and green quartzites and shales, the latter altered to dark grey and black phyllonites. The dolomites are on the other hand massive, somewhat boudin-like and consist of small to medium sized and strikingly isometrical dolomite crystals. West of this faulted area along the Uri Chu, a tributary to the Manas river (Dangme Chu) the Geological Survey has investigated larger gypsum deposits related to phyllitic slates, limestones and calcareous slates. *Nautiyal* places these outcrops into his Shumar Series and correlates it with the Dalings (*Nautiyal et al.*, 1964). Northerly dipping sericitic phyllites and quartzites seem to be thrust over the gypsiferous and calcareous horizons. I prefer to correlate this section with part of the Baxas, while *Jangpangi* (1974) places them within his Shumar nappe.

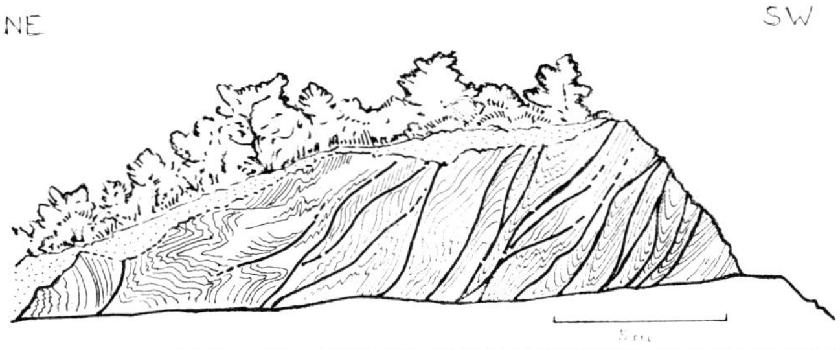

Fig. 12. Highly sheared slaty phyllites and phyllonites in the thrusted foothills of SW Bhutan, above Phuntsholing.

D.4. The Daling-Shumar Group

Apart from the Baxa carbonate rocks and associated shales, quartzites and iron-stones, the Bhutan Himalayan contains a most complex section of phyllitic shales, slates and quartzites. Lateral facies changes and a most severe tectonic overprint resulted in a marked disharmonic folding and an irregular cleavage system with smaller imbrications and larger thrust-zones. An incipient metamorphism increases from the structurally lower sections to the higher ones. All these facts characterise one of the most complex sedimentary groups of the whole Lower Himalaya. The local names and the contrasting age assignments of this practically nonfossiliferous sequence clearly reflect the uncertainty with which the Himalayan geologist is faced. For this still problematic belt I like to extend into the Bhutanese Lower Himalaya the wellknown name of Daling since direct correlation for at least part of this sequence has been established (*Sengupta* and *Raina*, 1978). In order to indicate the more quartzitic facies well developed in eastern Bhutan I like to add the term Shumar, introduced by *Nautiyal et al.* (1964) and further discussed by *Jangpangi* (1974, 1978) and speak about the Daling-Shumar Group. This group should be separated from the Baxa and placed, also according to the above mentioned authors, into the late Precambrian. *Nautiyal et al.*, (1964) have introduced a Phuntsholing and a Samchi Series for these sedimentary sequences. However, part of the Phuntsholing section could belong to the Lower Baxa and furthermore this area with its strong structural complications is rather an unfortunate type locality. The Samchi Series consists of phyllites and quartzites and is regarded by *Nautiyal* as an equivalent of the Daling. It seems therefore unnecessary to introduce here a new name. The Daling-Shumar Group covers the largest part of the Lower Himalaya of Bhutan, following with variable extensions just S of the large crystalline thrustmass.

In SW Bhutan good exposures, though obviously tectonized, were observed along the Phuntsholing road while this highway was under construction and the outcrops fresh (fig. 11). The hills just above the flood plain expose highly sheared, contorted and faulted slaty phyllites with silty bands, mostly dark grey to greenish grey (fig. 12). Phyllonitic horizons are nearly black and often strongly slickensided. A strong pressure cleavage and minute drag folds are the effect of a strong tectonization (phot. 4). The

Phot. 4. Thin section of disharmonically folded and sheared Daling phyllonites. Phuntsholing section.

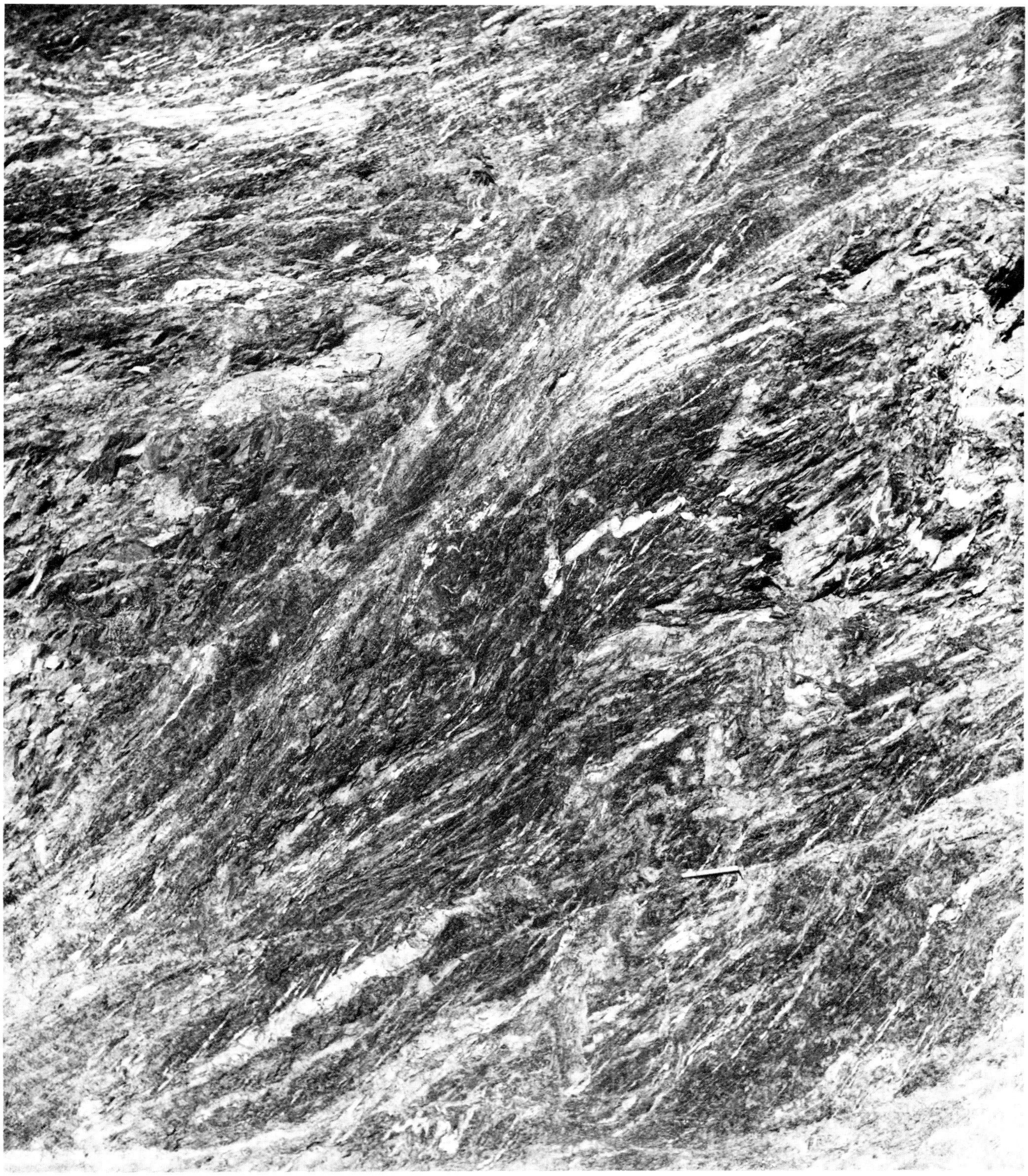

Phot. 5. Strongly foldet, sheared and locally thrusted phyllitic Dalings with disrupted thin quartzites. Phuntsholing section.

larger folds are overturned towards the southwest, and their near-horizontal fold axes strike NW–SW. Thin, finegrained quartzite layers which are highly fractured and disconnected within the sheared phyllitic masses occur locally. Larger thrust zones are invariably steep, dipping to the NW (phot. 5). All the evidence indicates a most severe tectonization which may be related to the thrust contact of the Baxas with the Daling slates and the vicinity of the Main Boundary Thrust.

Further northeast, after a tectonic wedge reminiscent of the Lower Baxas, follows a chaotic succession of faulted and thrusted slates and mostly white quartzites, the former intensely sub-folded, the

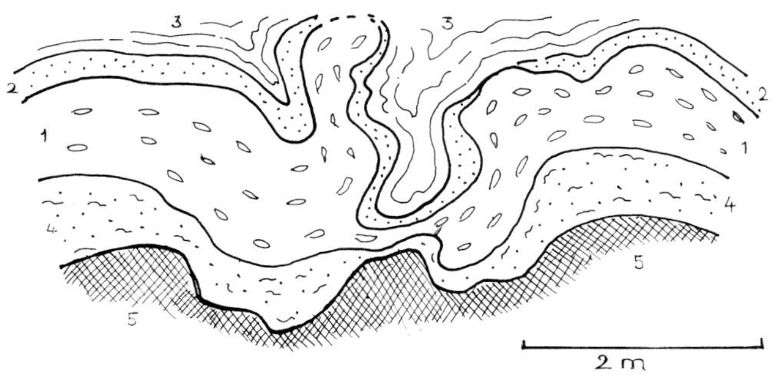

Fig. 13. Flow folding of biotite augengneiss (*1*) quartzites (*2*) phyllites (*3*) quartzitic schists (*4*) and amphibolite (*5*). Augengneiss belt below Chasilaka thrust. Axial plane vertical to steeply S hading.

latter fractured but more regularly bedded. Several steep thrusts can be inferred, but their magnitude is unknown since in this complicated section the stratigraphical sequence is incomplete. Intercalated phyllites resemble the sheared section at the base of the foothills. Along zones of disturbance we note the first dykes of fine-grained diabases and amphibolites, increasing somewhat in abundance towards the NE.

Towards Kamji the quartzites increase in number. They are white, fine-grained, thin to thick bedded with sericite partings. The intervening phyllites are more sericitized. Basic sills are still intercalated. One sill, about 100 m thick, consists of subeuhedral prism and laths of pale green actinolitic hornblende within a fine grained matrix of plagioclase, clinozoisite and biotite. Interspersed in this matrix is skeletal black ore together with apatite and biotite. No ophitic texture was seen. Unfortunately no unmistakable indication for bottom and top determination in the quartzitic horizons was noted. Some cross-bedding may indicate overturning of the Kamji quartzites, but the evidence is not too conclusive. Above and to the NE of Kamji, subfolding and shearing is again most intense in the argillaceous sections, now gradually becoming more metamorphosed. Sericite schists are formed, rich in irregular quartz veins. Within these tectonized sericite schists lies a biotite-augengneiss, in itself subfolded and sheared (fig. 13). The augen consist of perthitic orthoclase with myrmekitic rims surrounded by a mosaic of small quartz grains, some plagioclase, small garnets, larger biotites and some muscovites with sphene and epidote. Above and below the folded augengneiss are sheared micaceous quartzites, together with an amphibolized dolerite. These sills are common in the Phuntsholing section (fig. 11). The gneiss seems tectonically emplaced. It differs, however, from the granitic gneisses following higher up in the section. It could be compared to the various gneiss lenses within the Shumar formation described by *Jangpangi* (1974) which are not related to the main crystalline thrust. Similar gneisses were observed between Dalings and Darjeeling gneisses in eastern Sikkim by *Sengupta* and *Raina* (1978). Sericite schists above the augengneiss lamellae show a gradual increase of biotite together with the formation of larger muscovites. Sericitic quartzite layers are still present. Finally still intensely folded biotite-muscovite schists occur, already reflecting a higher grade metamorphism. They are overlain conformably, though with disharmonic contacts, by well- to thick-bedded biotite-granite gneisses, which lead to the large saucer-like crystalline mass of Chasilaka. In spite of intense shearing and sub-folding of the underlying meso-metamorphic schists it is difficult to trace a regional tectonic contact in the field between the argillageous rocks and the overlying granite gneisses.

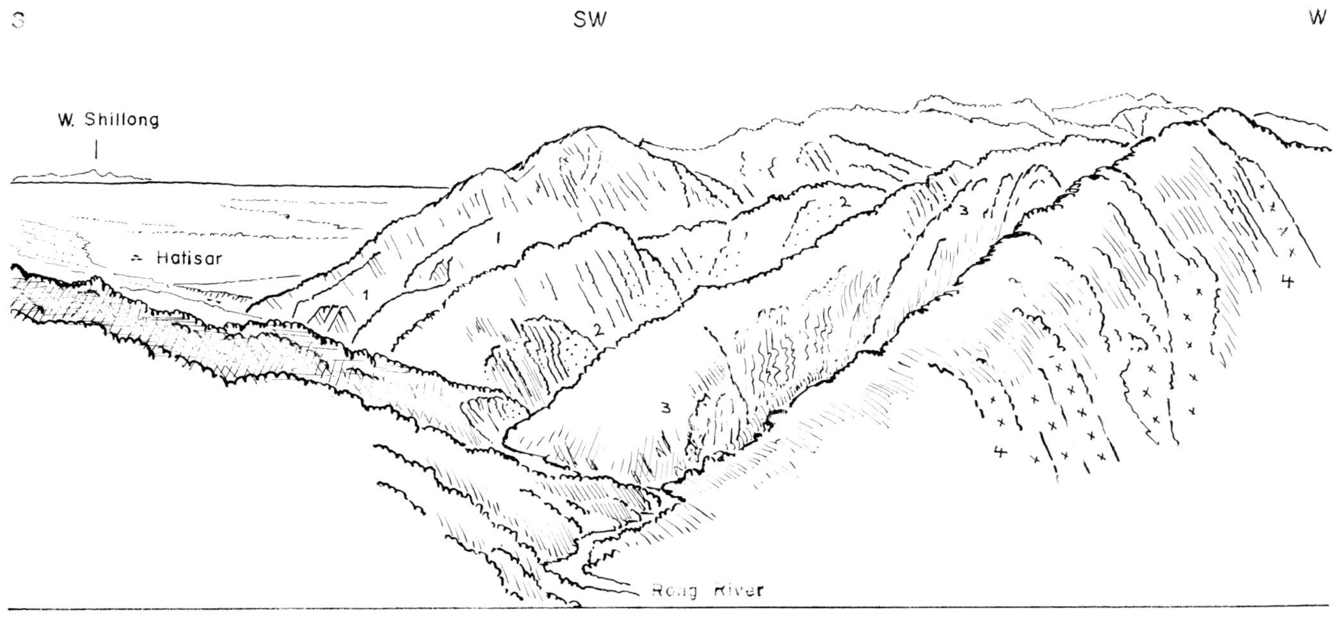

Fig. 14. The foothills of Hatisar. View into the Brahmaputra plain with the western Shillong shield rocks at he horizon. *1* phyllites, quartzites and quartz conglomerates, *2* main quartzites, *3* sericite schists, *4* Sure gneisses.

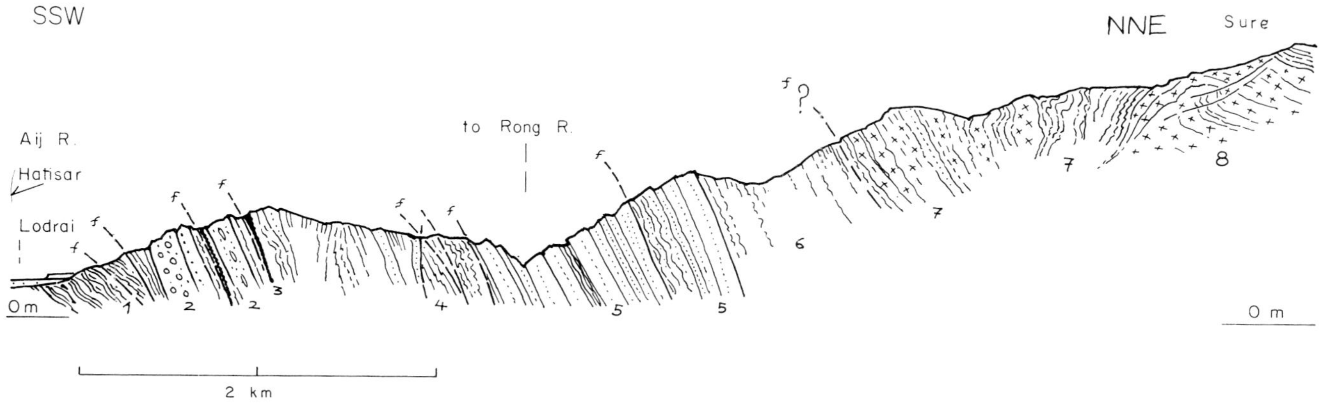

Fig. 15. Section across the foothills between Hatisar and Sure, S central Bhutan.
1 highly sheared green and violet phyllites,
2 red and green quartz conglomerates,
3 black graphite schists, mylonite zone,
4 green phyllites, Daling type,
5 platy, finegrained white qaurtzites,
6 sericite-biotite schists,
7 augengneiss, banded gneiss, sericite quartzites biotite schists,
8 biotite granite gneiss with garnets (Sure gneiss).

In southcentral Bhutan, a section reminiscent of the Daling-Shumar Group is well exposed along the Rong river valley north of the Hatisar (Geylegphu) embayment (fig. 14). Here again good outcrops were exposed during the road constructions to Tongsa. They are shown on a tentative section from Lodrai in the S to Sure (Soraya) in the NE (fig. 15). The sediments, corresponding to the Daling rocks, form a narrow and steep belt, with a rather constant abnormal strike direction from NW to SE. It is therefore questionable if these foothill rocks can be traced into the already described section north of Phuntsholing in SW Bhutan.

Highly sheared greenish and dark violet phyllites form the lowest foothill outcrops, but are covered by terraces of the Ai river near the Lodrai road camp. They dip towards NE and are overlain by conspicuous quartzitic comglomerates. Pink and violet, white and dark grey quartz and quartzites form pebbles varying from 1 to 5 cm. They are sub-rounded and covered by sericitic films. The matrix is coarse, gritty and badly sorted and rich in sericite and quartz. Northwards the conglomerates become intensely sheared and the quartz pebbles considerably stretched. They are bordered by conspicuous graphitic mylonitic shale zones. These conglomerates form a tectonic wedge and are unlike normal Dalings. They may represent a special facies within the Lower Baxas, since carbonate Baxas lenses are present W and E of the Hatisar embayment. With steep northeastern and locally vertical dips follow greenish phyllites and thin quartzite bands. Some of the green Daling-like phyllites can assume a real flowing pattern in the highly tectonized zones. These are particularly common near the thick, fine-grained quartzites of the Phuntsholing section, but here they are considerably thicker. Usually very well bedded with a quite constant dip of about 50–60° to the NE, they contain intercalated horizons of phyllites surrounding strongly boudinaged quartzite and vein quartz lenses (phot. 6). The quartzites are cut by a conspicuous cleavage system intersecting the steep bedding with mostly flat and some steeply southwards-dipping joints (fig. 16). Locally largescale pinch-outs, accentuated by some bedding cleavage, somewhat resemble large cross-bedding. A dubious reversed sequence is indicated in this particular section (phot. 7). The main quartzites are over 1000 m thick and form the very steep ridges in this jungled region.

With a rather sharp break, the main quartzites are overlain by sericite schists, in which biotite increases gradually until real biotite schists are met. They still contain some thin sericite quartzite horizons which remain surprisingly constant in spite of the change in metamorphic grade. Throughout the section the boitite schists are highly disturbed, particularly towards the first gneissic intercalation. Here again, a major disturbance may be present, but cannot be clearly seen in the field. The first gneisses are finely striated and banded biotite-muscovite gneisses, and certain horizons of typical augengneisses are found. This whole gneiss mass, with in-

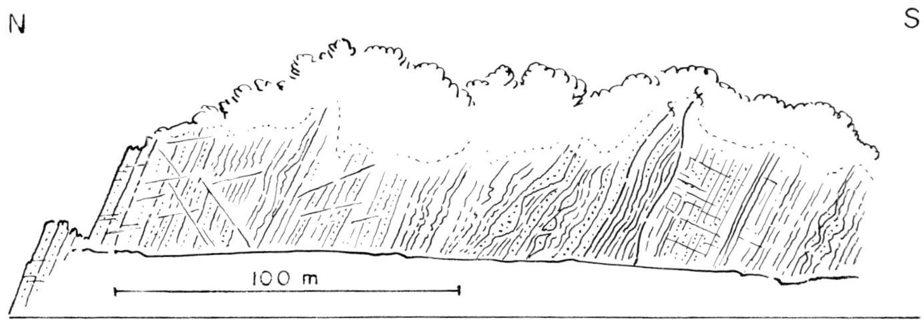

Fig. 16. Fractured and boudinaged quartzites and sheared phyllites between Hatisar and Sure. Corresponds to nr.5 in fig.15.

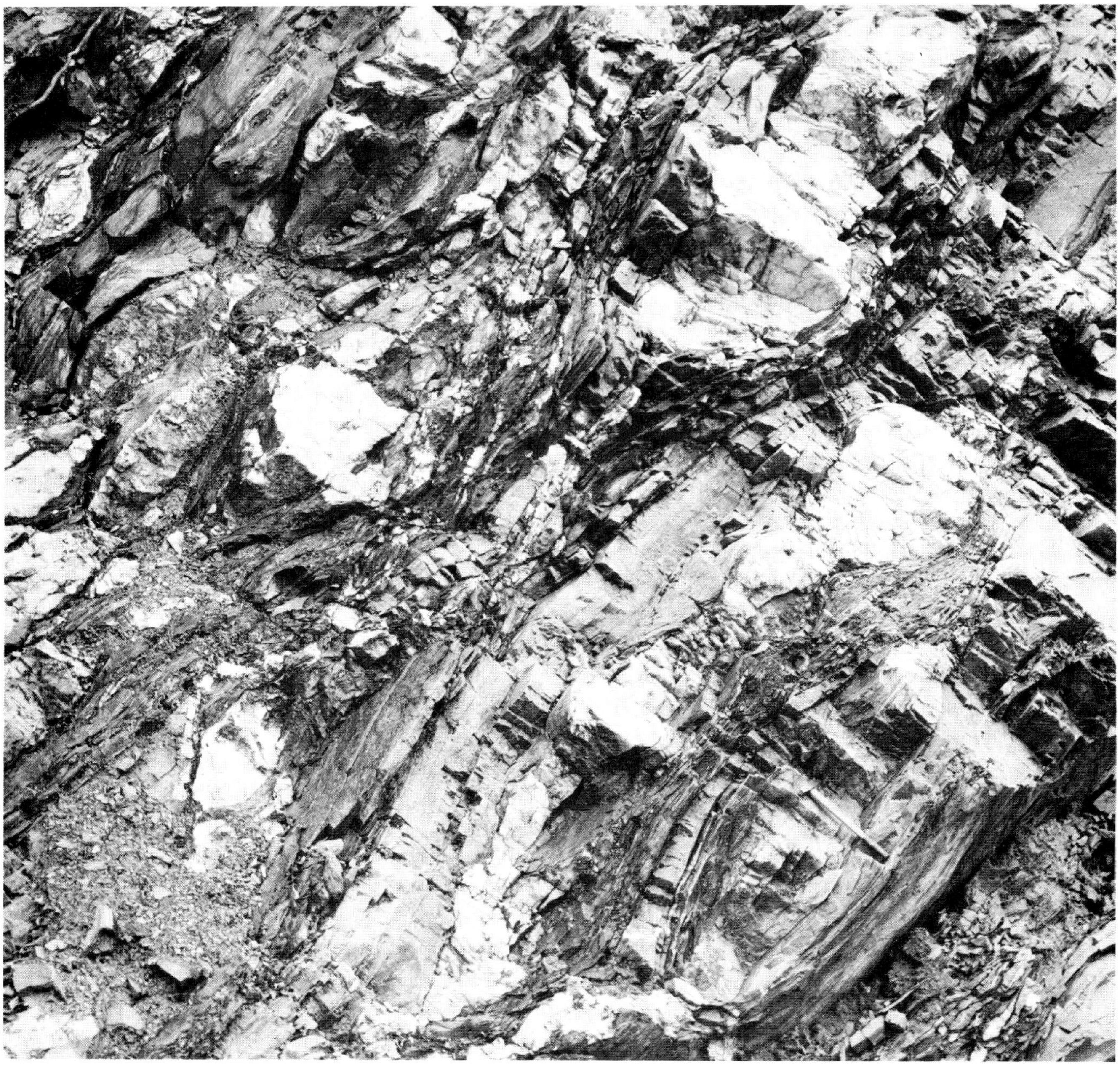

Phot. 6. Boudinaged quartzites and quartz veins, surrounded by phyllites and slates. Hatisar-Sure section.

tercalated biotite schists and some quartzite, will be discussed separately. It may correspond to the Chasilaka gneisses covering the Phuntsholing section, except that in our present area, the tectonic style of the crystalline mass is much more complicated and its overall dimensions are considerably smaller (fig. 15).

The thickest Daling-Shumar section is met along the Kuru Chu. Here the river follows the northwards directed half window or spur of the Shumar formations below the main crystalline thrust. It forms a complicated antiform, striking to the N and is cut, particularly towards the S, by conspicuous W to E directed fault and thrust zones, outlined on the map of *Jangpangi* (1974). Less disturbed sections were observed by the same author at Shumar and in the foothills to the east of the Tashigang road. He describes phyllites, phyllitic schists, thick bedded and flaggy quartzites, grey slates and thin epidioritic sills. He also includes gypsum beds, thin limestones and carbonaceous slates which may, as already mentioned, belong to Baxa type imbrications. The latter frequently outcrop as small lenses along the main fault and thrust zones. Between phyllites and quartzites facies changes are striking, and the latter can dominate over large sections.

Problematic are the newly introduced Thungsing quartzites of *Jangpangi* (1974) in the central part of southeastern Bhutan. They are in tectonic contact with the Baxa dolomites along the Tashigang road but the author reports interbedding with the Diuri boulder slates. He also compares them with the

Phot. 7. Highly cleavaged, well bedded quartzites with larger pinch-outs and topped by more slaty quartzites. Hatisar-Sure section.

quartzitic sandstones of the Gondwanas. Based on my own obsevations I doubt this correlation and I cannot see any difference between the white feldspathic quartzites of Thungsing N of Narphong (Tashigang road) and the well developed quartzites of the Shumar formation. The quartzites interbedded with the basal Diuri boulder slates may be some different horizons, since the normal contact is a clear transgression with a basal conglomerate.

The various rock types in the Upper Shumar section below the crystalline thrust at Barsong along the Tashigang road clearly show how complex these formations can be with several intercalations of a very doubtful provenence. The general situation of the Daling-Shumar Group is shown in figure 128 including a composite section from S of Tashigang to the southern foothills at Samdruk-Jongkhar. The northern border is rather sharp and coincides with the overthrust of the crystalline thrust sheet of Tashigang. The southern border is marked by the faulted transgression of the Diuri formation. Below the northern thrust contact, several imbrications expose a somewhat different lithology. A further structural anomaly occurs north of Narphong where the Baxa type dolomites outcrop.

The bulk of the formation is the quartzites which outcrop as well bedded white to greenish grey, fine to coarse grained platy quartzites grading into quite thick bedded layers. Thin bedded horizons have a white and grey banding. In zones where the argillaceous facies predominates the quartzites are boudinaged similar to the outcrops along the

Hatisar-Sure section. This and a frequently observed strong N and NW directed striation along bedding planes indicate marked stress effects related to pronounced disharmonic movements. The stress effect is less visible on thick bedded coarse grained quartzites. Regularly distributed quartz grains show no preferred crystallographic orientation just below the main thrust along the Tashigang road, but a strong undulate extinction. The greenish colour is caused by some chloritized biotites.

In the N, below the main crystalline thrust, quartzitic biotite-sericite schists prevail with fresh unaltered biotites and a polygonal type of quartz without any mortar structure. Two thin marble layers are intercalated, one fine grained and white and grey banded, the other coarser and white. Some of the bandig is underlined by fine quartz stringers. In spite of the fresh biotite content of the enclosing schist no reaction between calcite and quartz grains has been noted. Within the schists below the crystalline thrust one can also recongnize more lenticular layers with biotite, quartz, sericite and larger, highly sheared calcite crystals, together with pyrite and graphitic matter. These rocks may represent a highly sheared quartz-carbonate conglomerate with an argillaceous matrix. Below follow quartzitic schists and phyllites with a very pronounced crenulation type subfolding with eastwards dipping fold axis and vertical axial planes, contrasting with the generally north dipping more competent beds. In this section is also an intercalation of strikingly parallel and well bedded (1 cm thick beds) siliceous, light and dark grey banded carbonate rocks. This striking sequence, less than 100 m thick, contrasts with the contorted and sheared phyllites. The dark grey calcareous layers contain organic matter. A careful investigation by *Nicola Pantic* produced some unexpected results, which have been published in a separate paper (*Pantic et al.*, 1981).

The dark banded calcareous slates contain determinable palynomorphs. Apart of some acritarchs and fragments of chitinozoa suggesting a reworked older Paleozoic (Silurian-Devonian) they produced Jurassic palynomorphs which allow a quite precise dating. Dinocysts are of special importance for the biostratigraphic interpretation. They indicate the lower part of the Upper Jurassic (Oxfordian). The strong predominance of dinoflagellate cysts in relation to pollen and spores suggests an offshore area.

The presence of an Oxfordian florula just below the Main Central Thrust and above a thick, mostly Precambrian section, the Daling-Shumar group, which represents a northern facies of the Indian shield, is of particular interest. It differs from the Tethyan facies in the N as well as from the known Gondwana type sediments in the S. We may have here the rare fact that sediments with fossils from an intermediate facies (between Gondwana and Tethys) which normally are covered by the large crystalline thrusts have been transported to the S along the Main Central Thrust and locally preserved along the front.

The reworked Lower Paleozoic fossils suggest erosion and subsequent reworking during late Paleozoic epirogenetic movements. The nearest Paleozoic sediments outcrop in the Tang Chu basin, described below.

If we consider that this formation has been overthrust at least 80 km by an over 15 km thick crystalline body it is most surprising that some of the Jurassic fossils are so well preserved, some still showing the delicate membrane. Most of the stress from the thrusting must have been absorbed by the few meters of dislocated schists above the well bedded sediments. Thrusting effect and mechanism is still little understood!

I tentatively call this section the Barsong unit, named after a locality just below the main thrust along the Tashigang road (fig. 128).

South of Barsong outcrops a southwards dipping limb of the thick quartzite and shale layers of the Daling-Shumar group which form a wider synclinal basin with an eastwards rising axis. The southern flank can be followed until the above mentioned contact with the Diuri formation where it is disrupted by the tectonic wedge of Baxa dolomites and quartzites. The minimal thickness of the Daling-Shumar Group amounts to approximately 5000 m if this flank represents a normal sequence. The resulting large volume of detrital quartz found in this section is surprising , particularly since most of the quartzites are pure and arkosic beds rare. A direct origin from a crystalline basement is unlikely. Unfortunately, sedimentary features which would assist to determine top and bottom are inconclusive. A vague crossbedding pattern and some ripplemarks have been observed. Well developed is however a strong lineation and intense subfolding which affected the Shumar as well as the Diuri deposits with the same intensity. Most of the lineation is E–W oriented with dips of 10° to 30° in both directions, suggesting a gentle subsequent undulation. It may be related to a mega folding phase. Only locally one can distinguish a sharp northwards directed lineation and small amplitude folding which seems to overprint the general E–W directed structures (fig. 10), and coincides with the direction of

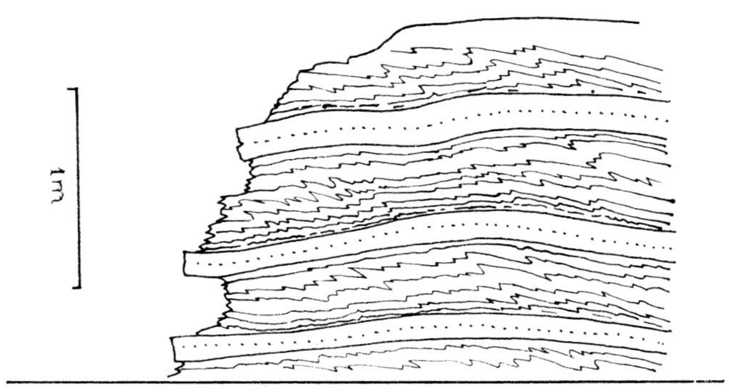

Fig. 17. Drag-folds in sericite schists between quartzite layers. S dip 285°/30° and foldaxis 345°/15°. Metasediments of the northern Kuru Chu-Shumar Spur.

pebble stretching. In spite of the strong tectonization of this section all minerals are fresh and, judging from the composition of the other schists, no sign of diaphtorism can be noted. Recrystallization must have occurred mostly after the structural overprint and the metamorphic grade seems to decrease downwards in the sequence and towards the S, similar to the Phuntsholing section, a most important regional fact in the Lower Himalaya.

The thrust discussed at Barsong can be followed in a NW direction into the Kuru Chu valley, S of Lhüntse Dzong, where the northern plunge of the Shumar-Kuru Chu spur outcrops. Here, the otherwise sharply outlined thrust is practically lost since the metamorphism of the underlying Daling-Shumar Group has increased in grade considerably. We thus place this contact arbitrarily below the first thick granite gneiss layer, about 4 km to the SE of Lhüntse Dzong. Here on the western side of the valley outcrops a biotite-muscovite-garnet bearing gneiss intruded by tourmaline leucogranite. The latter also contains minor muscovite and garnet. On the eastern side of the Kuru Chu valley, the gneisses are very well bedded and alternate with biotite augengneisses and banded two mica garnet gneisses with rotated garnets.

All these gneisses of the northern main crystalline unit are thrust and draped around the north plunging antiform of the Shumar rocks. Below the gneisses, (below the thrust), muscovite quartzites outcrop. These are often well bedded and alternate with two mica schists which in turn are intensely subfolded (drag folds) between the smooth quartzites (fig. 17). They overly more than 100 m of white and gray banded calcite marbles. The white bands are coarser than the grey ones, not unlike some layers of the Paro marbles. They contain some biotite but no phlogopite. Below the marbles follow thick and thinner bedded quartzites. Here some crossbedding is still visible, suggesting a normal position of the layer. The thin bedded quartzites also alternate with mica schists with a very pronounced disharmonic folding (fig. 18). Some of the quartzites have folds reminiscent of flow folding (fig. 19). All the fold axes plunge towards the N about 10–25° and are parallel with the megafold of the N plunging Shumar antiform. The regional NW and NE dips of the antiform vary between 25° and 35° respectively. To the S the metamorphic grade decreases and gradually more phyllitic rocks occur.

Of special interest are the newly discovered granit gneisses within the Daling-Shumar Group of southeasten Bhutan (*Jangpangi* 1974). Known are now 5 larger bodies extending for several km and related to anticlinal uplifts (Gachang granite gneiss in the Kuru Chu spur) or to internal fault and thrust zones, the largest of which are Jangthi and Ani-Uni bodies, which both continue eastwards into Arunachal Pradesh (NEFA). They consist of muscovite biotite granites and granite gneisses with a more or less porphyritic texture. Phenocrysts of perthitic orthoclase and quartz are set in a groundmass of plagioclase, muscovite, biotite and sericite. Only the larger bodies are massiv but change along their borders from augengneiss to blastomylonites and phyllonites. The contacts are tectonized and no contact metamorphism was observed. Originally the granites may have been intrusive into the Shumar sediments. They are different from the granitic rocks of the crystalline thrust sheets which are migmatitic with pronounced granitization. As already alluded to, they could be correlated, similar to the augengneiss in the upper Phuntsholing section, to the gneisses in eastern Sikkim below the main crysralline thrust. Similar gneisses have been reported from the Kameng district (NEFA) adjoining Bhutan to the E. (*Das et al.*, 1975; *Verma* and *Tandom*, 1976). They are intercalated in the lower part of the Bomdila group of metasediments and occur as biotite augengneisses, locally strongly foliated and mylonitic. Crystalline intercalations within the Bomdila metasediments are however more diversified than in the Shumar region of SE Bhutan. This seems also the case for the sediments of the Bomdila group which, on general lines, are the eastern aequivalent of the Daling-Shumar group.

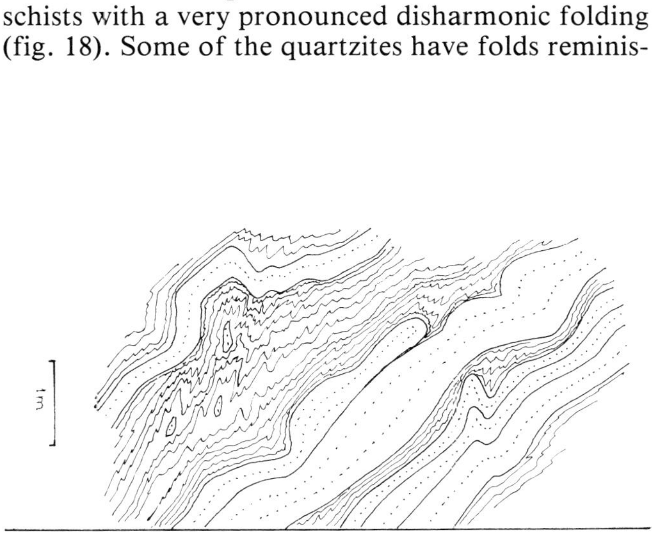

Fig. 18. Deformed metasediments of the northernmost Kuru Chu Shumar spur: Disharmonic folding between quartzites and sericite schists. Foldaxis in quartzite 350/30, schists10/25.

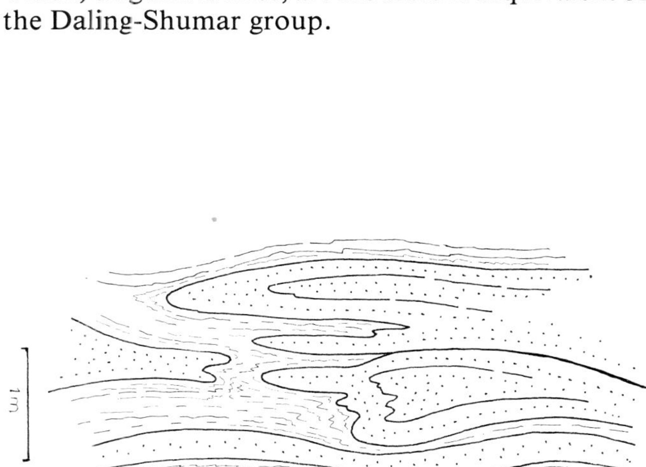

Fig. 19. Deformed metasediments of the northernmost Kuru Chu Shumar spur: Flow folding in quartzites with garnet-sericite schists. Foldaxis horizontal striking to 20°.

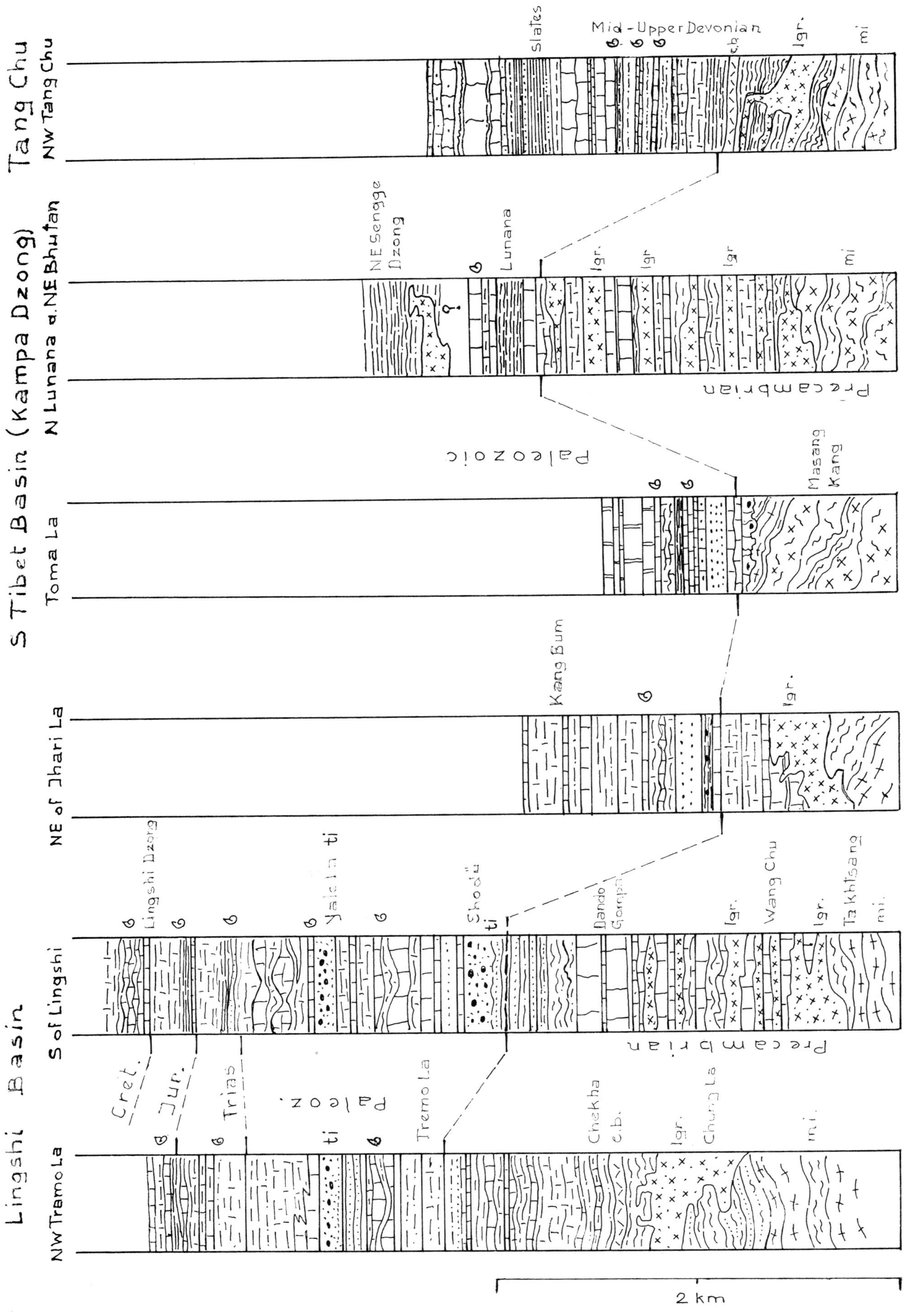
Tang Chu
NW Tang Chu
slates
Mid-Upper Devonian
c.b.
lgr.
mi
S Tibet Basin (Kampa Dzong)
N Lunana a. NE Bhutan
NE Sengge Dzong
Lunana
lgr.
lgr
lgr
mi
Precambrian
Paleozoic
Toma La
Masang Kang
NE of Jhari La
Kang Bum
lgr.
Lingshi Basin
S of Lingshi
Lingshi Dzong
ti
Shodü ti
Dando Gompa
lgr.
Wang Chu
lgr.
Takhtsang
mi.
Precambrian
Cret.
Jur.
Trias
Paleoz.
ti
Tremo La
Chekha c.b.
lgr.
Chung La
mi.
NW Tremo La
2 km

E. The Bhutan Tethys Himalaya

In the greater part of the Himalaya the Tethyan sediments form a normal cover above the crystalline belonging to the Main Central Thrust, the so-called Dalle du Tibet of *Lombard* (1958). Contrasting with the sharp tectonic base of this thrust mass, the top exposes a gradual and conformable contact with Tethyan sediments. They are mostly of platform type, ranging in age from the Cambrian to the Eocene. Regional metamorphism has reached various stratigraphic levels in various places within the Tethyan sections: Precambrian to Cambrian in the Spiti basin, Cambrian in the Kumaon Himalaya and Ordovician in the Nepal Himalaya and north of Chomolungma (Everest) in the wide Kampa Dzong basins (*Gansser*, 1974; *Mu et al.*, 1973). Only locally were younger beds involved, such as Trias and Lower Jurassic in southwestern Ladakh (*Honegger*, in press). Within the flysch sections of the northeastern Tethyan sediments does higher grade metamorphism affect the Middle Mesozoic (*Chang et al.*, 1977; *Hayden*, 1907).

The Tethyan sediments usually begin with a more or less fine detrital to argillaceous section. It straddles the Precambrian-Cambrian period in age and can reach over 5000 m in the Kumaon Himalaya. Part of this huge pile of sediments are calcareous, with calc-schists and marbles dominating in central Nepal Himalaya. The Tethyan sediments of Bhutan occur in basins and northern dipslopes: in the northwestern part as Lingshi basin, northern part as Toma La and Lunana belt, and in Central Bhutan as Tang Chu basin. They are always conformable above the main crystalline thrust masses (fig. 20).

E.1. The Lingshi basin

Here the Tethyan sediments form a well outlined basin of about 900 km² outcrop in the northwestern part of the country. This basin is dominated by the Lingshi Dzong which coincides with the youngest sediments (Cretaceous) and is thus given the name Lingshi basin. It connects with the Tethyan sediments of the Phari-Tuna region of South Tibet (*Hayden*, 1907) belonging to the wide Kampa Dzong basin of

◀ *Fig. 20.* The Tethyan sediments of the Bhutan Himalaya following above the Precambrian basement forming the main crystalline thrust sheet above the MCT. *mi* = migmatites, *lgr* = leucogranites, *cb* = zone of cross-biotites, *ti* = tilloid horizons. (For legend see also fig. 128.)

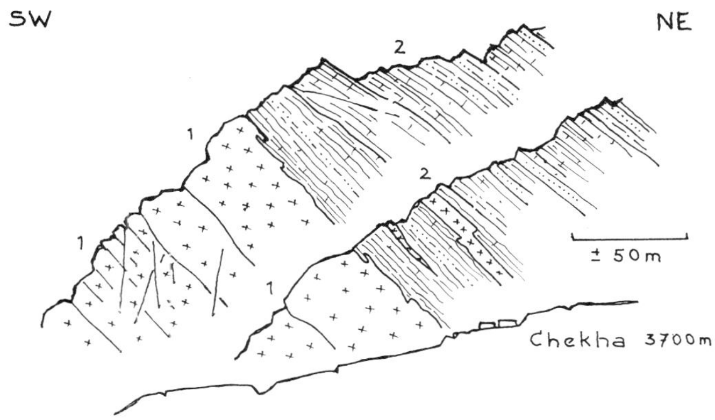

Fig. 21. Contact of leucogranite with Chekha formation at Chekha. *1* leucogranite of Chung La, locally fractured, *2* Chekha formation (quartzitic schists and calc-schists).

Tibet (*Gansser*, 1974). At present the connection is only visible over the Tremo La watershed and this only in the lowest, slightly metamorphosed sediments. Intrusions of young tourmaline granites have also disrupted the connection (see Geol. Map Pl. I).

The Tethyan sediments gradually develop from the top of the crystalline pile. We observe a gradual decrease of metamorphism within the argillaceous section. Here occur the typical biotite porphyroblast schists which we noted already in 1936 in the Garbyang section along the western border of Nepal (the Budhi schists of *Heim* and *Gansser*, 1939). They are identical in position and petrology and well exposed near Chekha, above the intrusive contact with the young tourmaline granites (fig. 21). Striking are the large porphyroblasts of cross-biotites with a brown pleochroism and sieve like inclusions of quartz (phot. 8). Some alignement of quartz inclusions seems to in-

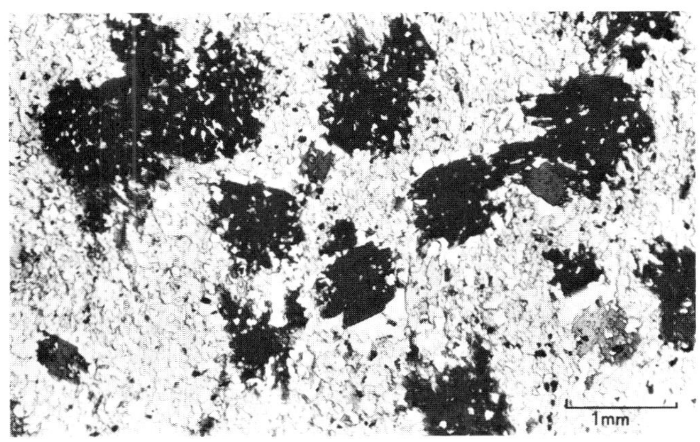

Phot. 8. Cross biotites with poikiloblastic inclusions of quartz. Base of Chekha formation.

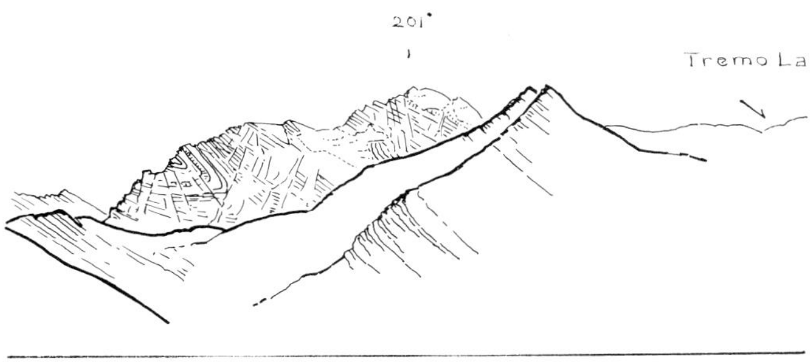

Fig. 22. The western continuation of the lower calc-schists and biotite schists of the basal Lingshi basin. Folds and regional NW dip in mountain group SW of Tremo La (Masang Kyungdu).

dicate a previous schistosity rotated up to 90° from the present foliation plane. The groundmass contains tourmaline and fine magnetite. The cross-biotites are well recognizable already in the field.

The number of porphyroblasts decreases upwards together with increasing thin and fine grained phyllitic muscovite bearing quartzites and calc-schists which can be followed up to the Tremo La, dipping westwards towards the Phari Dzong basin in Tibet. The calc-schists which dominate the section with some marbles and dolomites still contain some smaller biotites and are strongly lineated with well oriented calcite grains trending NE with 20° to 40° dip. This section has been called Chekha series by *Nautiyal et al.* (1964) and Chekha formation by - *Jangpangi* (1978). I would have preferred the name Tremo La formation, since Chekha is only a small military place not on the maps while Tremo La is the most important pass between Tibet and Bhutan. Southwestwards the Chekha formation can be followed into the high border range which lies in the source of the Ha river in W Bhutan (Masang Kyungdu, 5579 m, on the Indian map of Bhutan, 1972). The well bedded formations are clearly folded here (fig. 22). Eastwards the Chekha beds are cut by the young tourmaline granites of Chung La and the direct connection with the large Lingshi basin, of which they would form a normal base, is not visible (fig. 23). Southewards the Chekha facies at the base of the Lingshi basin is displayed in the over 6000 m high Chatorake mountain, which dominates the upper Paro Chu valley (fig. 24, 25).

Along the southern rim of the basin we find good contacts of the crystalline thrust sheet with the overlying formation S of Shödu and in the Wang Chu gorge. At Shödu, a small place N of Thimphu, we note that within the upper part of the Chekha calc-schist section the lime content has incraesed resulting in thick marble horizons. They are well exposed in the lake country S of Shödu and in the deep Wang Chu gorges to the E and SE of it. In the upper part the carbonate rocks are less recrystallized and we note some flaggy, finegrained limestones with irregular small patches and lenses of dolomite, the latter somewhat granular with rounded nobs which may even suggest some vague fossil remnants. Downwards in the section argillaceous bands are intercalated within the marbles, mostly as chlorite-sericite phyllites in the upper part and biotite schists in the lower sections, near the contact with the crystalline rocks. The biotite schists frequently show the typical, cross-oriented porphyroblastic biotites of the Budhi type, mentioned from the Chekha region. The actual contact consists of intensely subfolded biotite psammite gneisses and calc-silicates and marble bands (fig. 26). The psammite gneisses represent the upper part of the crystalline section. Tourmaline granite dykes and sills cut the contact zone and enter the lower marble horizons.

In the Wang Chu gorge the carbonates are very well exposed (fig. 27). Steep thrusting seems to double the thickness of this section which is normally over 1200 m thick. The regionally gently northwards dipping layers indicate the southern continuation of the structurally rather flattish Lingshi basin. Fracturing is well developed in the upper horizon, but downwards flow-folding begins in the marbles together with a marked increase in metamorphism.

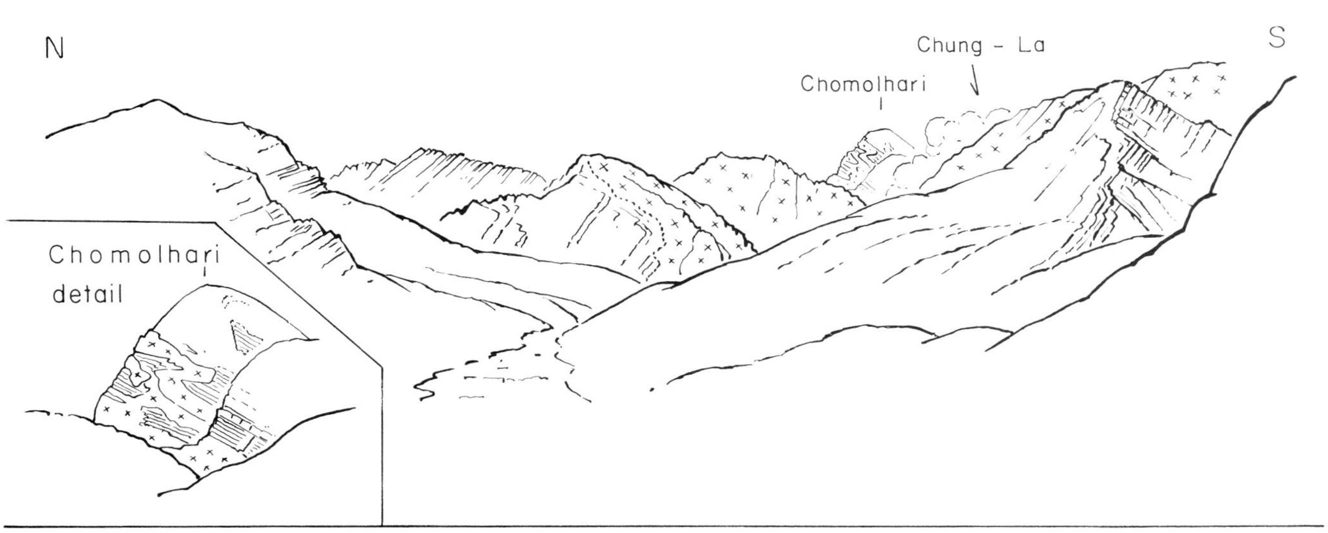

Fig. 23. The Chung La valley towards Chomolhari.
Calc-schists and phyllites (Chekha formation) intruded by the Chung La leucogranites.

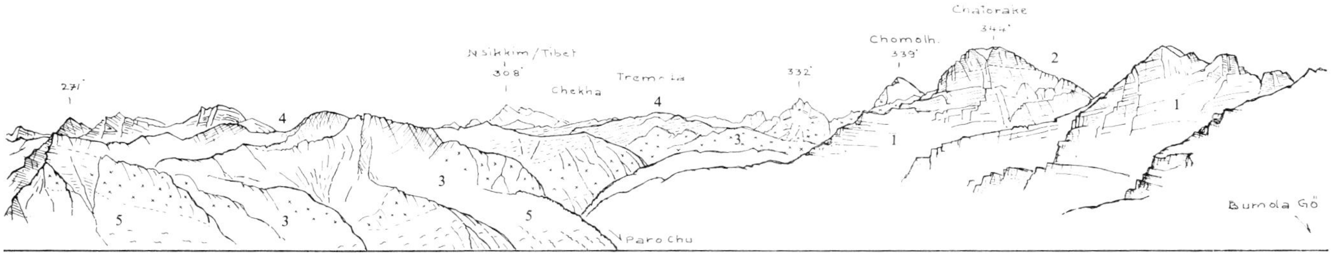

Fig. 24. Panorama of upper Paro Chu valley from peak 4050 m N of Bhumda Gömpa.
1 Well bedded muscovite-biotite gneisses, locally quartzitic with garnets and kyanite,
2 lower Lingsi sediments capping Chatorake peak,
3 tourmaline leucogranites of Chung La centre,
4 Tremo La formation,
5 migmatite gneisses of Takhtsang type in Paro Chu valley.

The fold axes conform to the general westwards dipping lineation. Sericite phyllites and phyllitic calc-schists form intercalations within the more compact calc-schists and marbles. Lamination, and in the marbles flow folding is specially intense just above Barshong Dzong. Further downstream banded calc-silicate marbles appear as deepest horizons with diopside, some plagioclase and biotite. Here we note the first lit par lit sill intrusions of fine-grained muscovite tourmaline granites, with tourmalines often forming radial nests. Calc-silicate marbles with biotite fels layers alternate with 20–50 m thick tourmaline granite sills for over 500 m. Dykes are very rare along this interval. Calc-silicate marbles and biotite fels are conformably overlying banded psammite gneisses which still contain some small calc-silicate bands. Over irregular more or less migmatitic granite gneiss we reach the typical Takhtsang crystalline section at Cäri Gömpa.

The same carbonate profiles are excellently displayed in the wild south wall of the Chomolhari range striking from the northeasten Lingshi basin to the SW of Chomolhari at Lamo La (fig. 116). The tourmaline granite of Chung La (ENE of Tremo La) profusely intrudes the Chomolhari marbles and schists, mostly in the form of larger sills. They gradually die out towards the NE, and there is an increase of the carbonaceous facies. These intrusions have overprinted the regional metamorphism by a pronounced but local contact metamorphism. Within the Chomolhari main group, argillaceous rocks are still present, with some biotite-sillimanite-garnet-gneisses and schists (fig. 28, 29). It is difficult to decide if these outcrops represent slivers of the older «basement» of the crystalline thrust sheet (Takhtsang type) or fully metamorphosed sedimentary cover. The limestones are altered to marbles and the impure thinner layers as well as the margins of the thicker beds to layered calc-silicates. They consist of calcite crystals, some quartz and basic plagioclase, biotite or phlogopite and diopside with locally scapolite. The spectacular alternations of tourmaline granite sills, biotite felses and marbles are visible throughout the 3000 m southwall of the Chomolhari range (phot. 9, 10). In the upper part the tourmaline granites have increasing abundance of xenoliths of biotite psammite gneisses and biotite felses, while the carbonate layers decrease (fig. 28, 29). Inspite of being largely sills, the intrusions often show discordant contacts to the banded marbles. The latter locally exhibit intense flow-folding with small but also large fold-amplitudes (phot. 11, 12).

To the northeast along the Chomolhari range, the intrusions decrease and the last sills are visible in the thick marble walls of the Dando Gömpa NE of Lingshi. At the same time the carbonate facies has considerably increased, witnessed by the impressive nearly vertical walls dominating the Dando and Dju

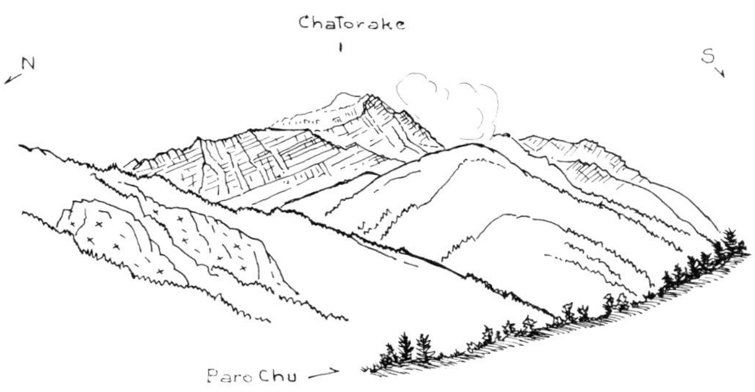

Fig. 25. Chatorake with southwards rising sediments of Lingshi basin. Base Takhtsang gneisses. Seen from above Gunisava, Paro Chu.

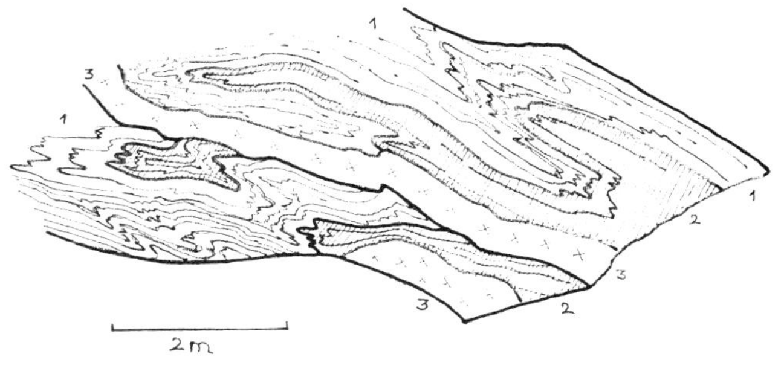

Fig. 26. Detail of intense flow folding of biotite-psammite gneiss and calc-silicate marble cut by post folding tourmaline granite aplite S of Tagetang (lake area NW Thimphu).
1 biotite-psammite gneiss, *2* calc-silicate marble, *3* tourmaline aplite granite. Foldaxis 265/15.

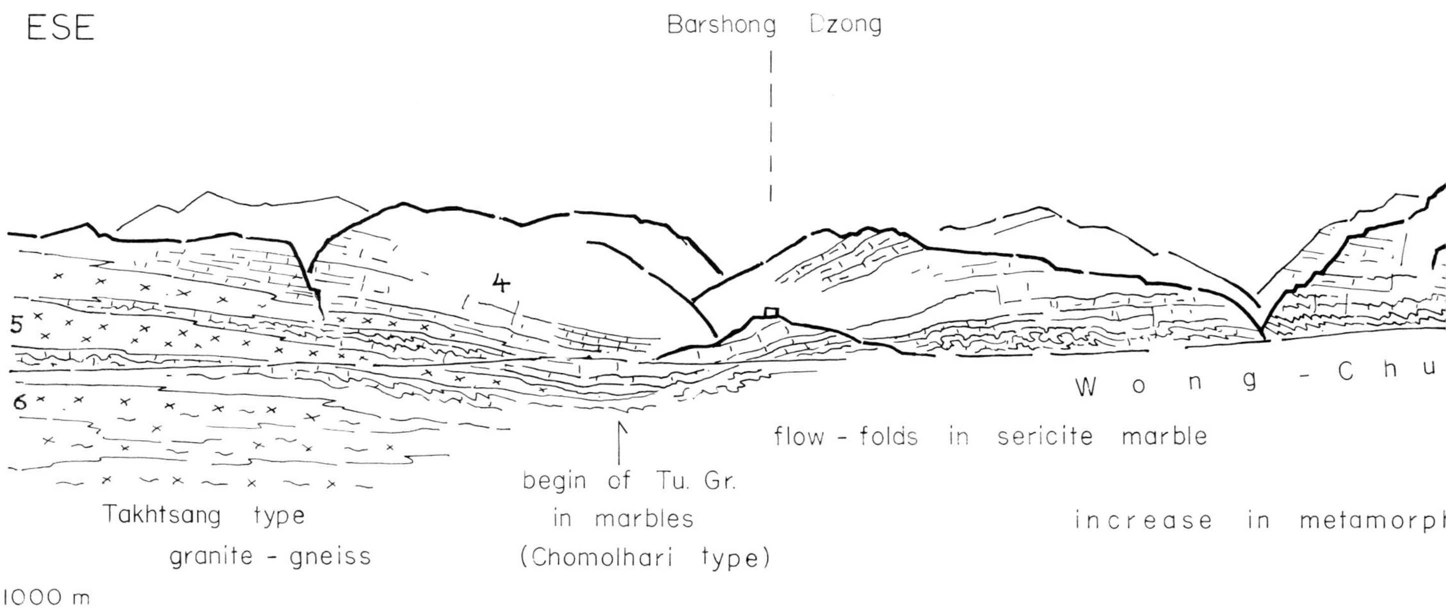

Fig. 27. Section along the Wang Chu valley (upper part),
1 limestone lenses and crinoidal calc-schists,
2 tilloid graywack zone of the Schodü area,
3 highly folded cherty quartzites and red slates,
4 marmorised calc-schists and marbles, often with flow structures,
5 leucogranites of Chomolhari type,
6 migmatitic gneisses of Takhtsang type.

Gömpas both built into their base. This characteristic increase in the carbonate facies of the Chekha formation as already observed in the Wang Chu gorge and NE of Chomolhari I would call the Dando Gömpa facies (fig. 30). Thick-bedded at the base it becomes thin-bedded and slaty towards the top. It is of special interest that where the Dando Gömpa facies is developed, carbonates are seen to conformably overly the crystalline rocks. This fact is also evident in the more northeastern sections of Bhutan, at the base of the Tethyan sediments. The contact of the Chomolhari-Dando Gömpa limestones with the normally overlying and younger Lingshi sediments is faulted and flexured (fig. 31). The Dando Gömpa limestones are bent to the SE and are subsequently cut by steep, southdipping faultzones (fig. 32, 33). The steepening SE flanks show small drag folds which are sharply cut by faults. Perpendicular to these structures are radial shear zones, along which

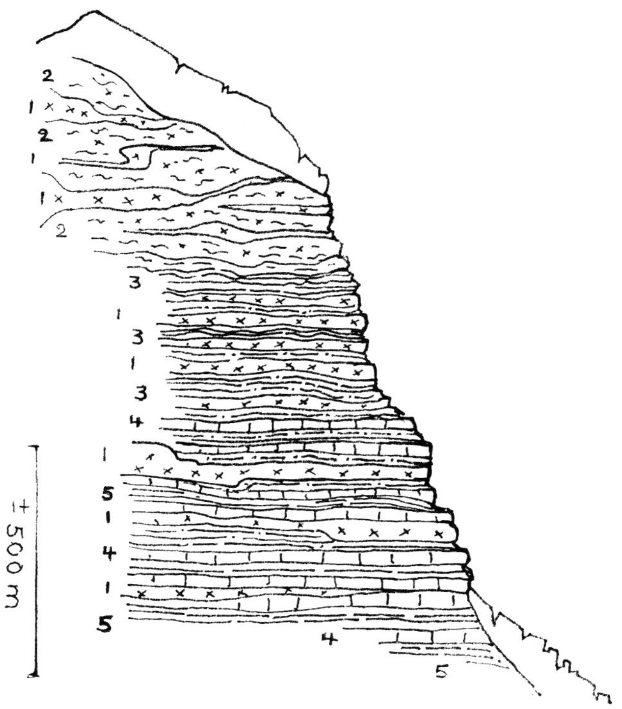

Fig. 28. Generalized section of SE wall of Chomolhari.
1 Tourmaline granite,
2 migmatitic gneisses - banded gneisses,
3 fine biotite-psammite gneisses and schists,
4 coarse marbles,
5 slaty marmorized limestone bands.

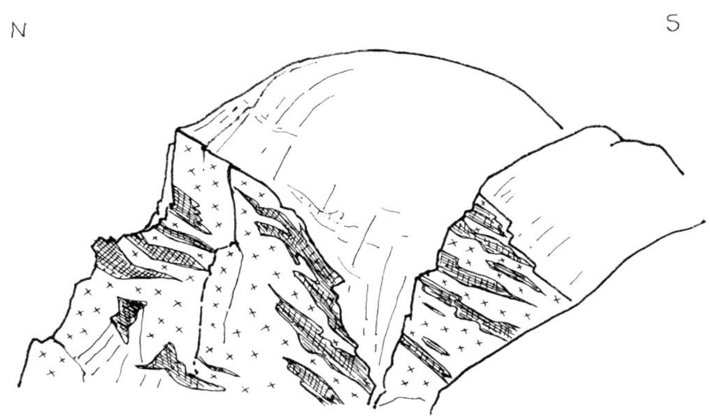

Fig. 29. The summit from Chomolhari. Tourmaline granite with xenoliths of biotite psammite gneiss and probable calc-silicate bands. (Sketched from photo of India Bhutan Chomolhari Expedition 1970 and own telephoto).

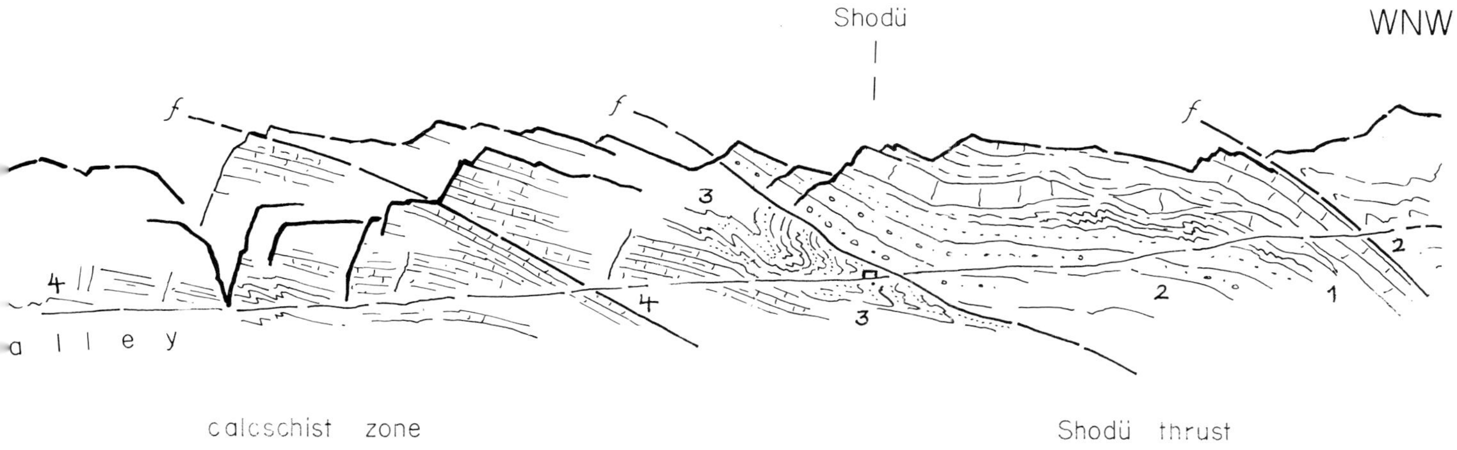

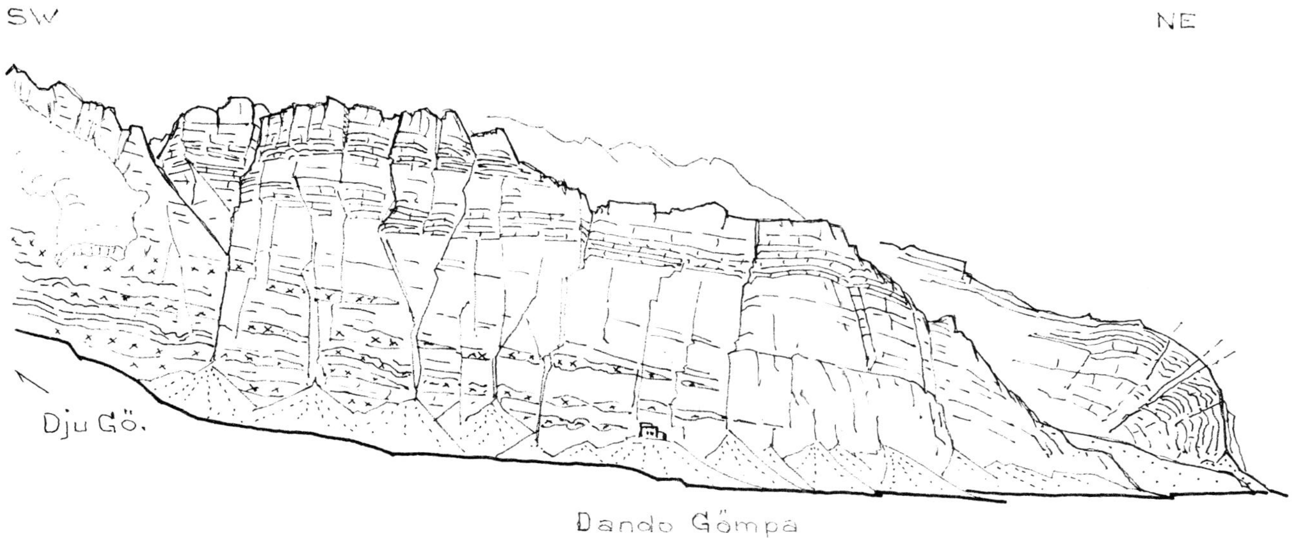

Fig. 30. Granite injections into Chomolhari marble zone SW of Lingshi. Base: thick bedded to massive marbles topped by thin bedded marbles and calc-silicates and followed by marmorized calc-schists and well bedded limestones. Decrease of metamorphism upwards and eastwards. This coincides with the westwards increasing intrusions of tourmaline granites which end at about Dando Gömpa. Note frontal flexure in the E, developing further northeastwards into the Lingshi fault zone, bordering the Lingshi basin in the NW (Fig. 32.).

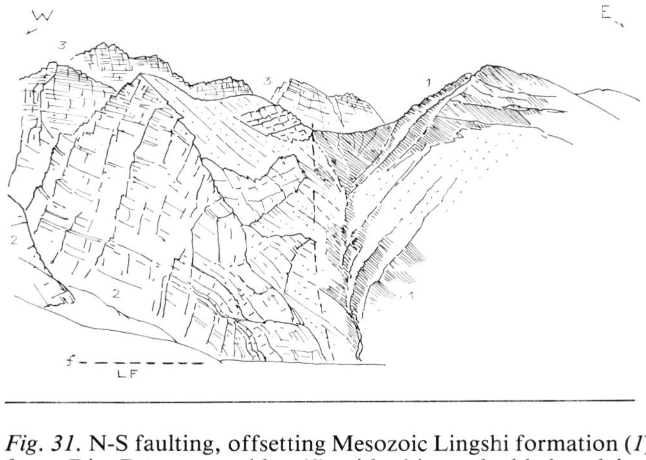

Fig. 31. N-S faulting, offsetting Mesozoic Lingshi formation (*1*) from Dju Dömpa marbles (*2*) with thinner bedded and less metamorphic top (*3*). View to the N, E of Lingshi.

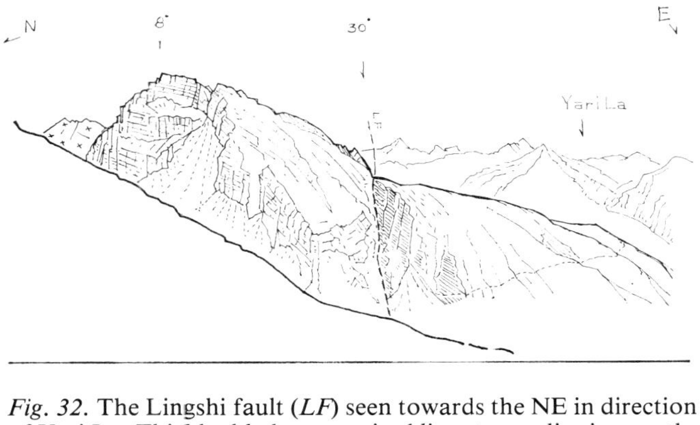

Fig. 32. The Lingshi fault (*LF*) seen towards the NE in direction of Yari La. Thickbedded marmorised limestones dipping southwards and faulted against Mesozoic black shales.

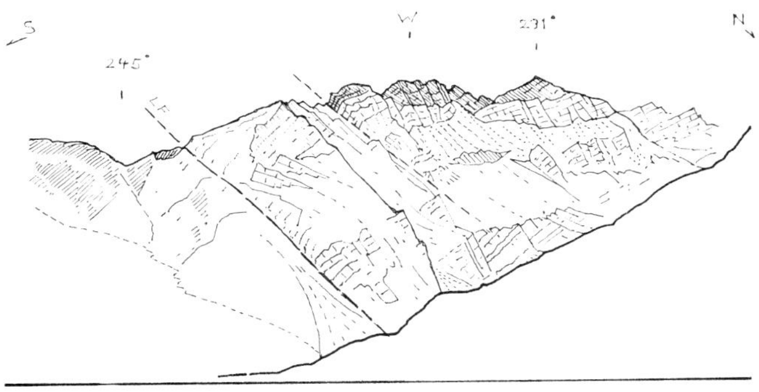

Fig. 33. The Lingshi fault (*LF*), seen towards W, NE of Chebetsa, Lingshi. Thick bedded marmorized limestone, topped by dark grey graywacky slates is faultet against black silty shales in the S. Mesozoic of Lingshi basin.

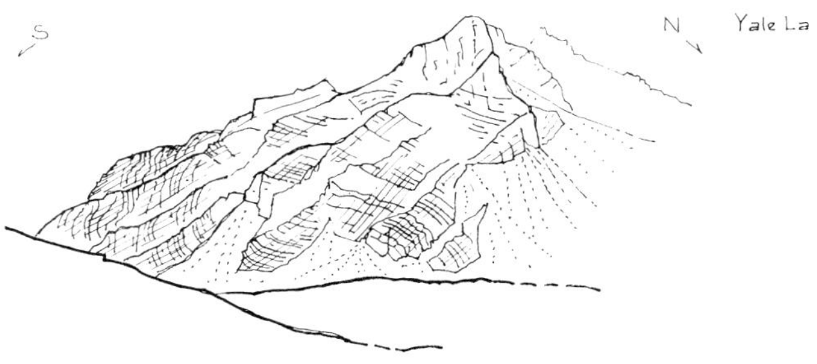

Fig. 35. Limestones N of Shodü, on way to Yale La. View towards W. Strong parallel cleavage cutting folded limestones bands.

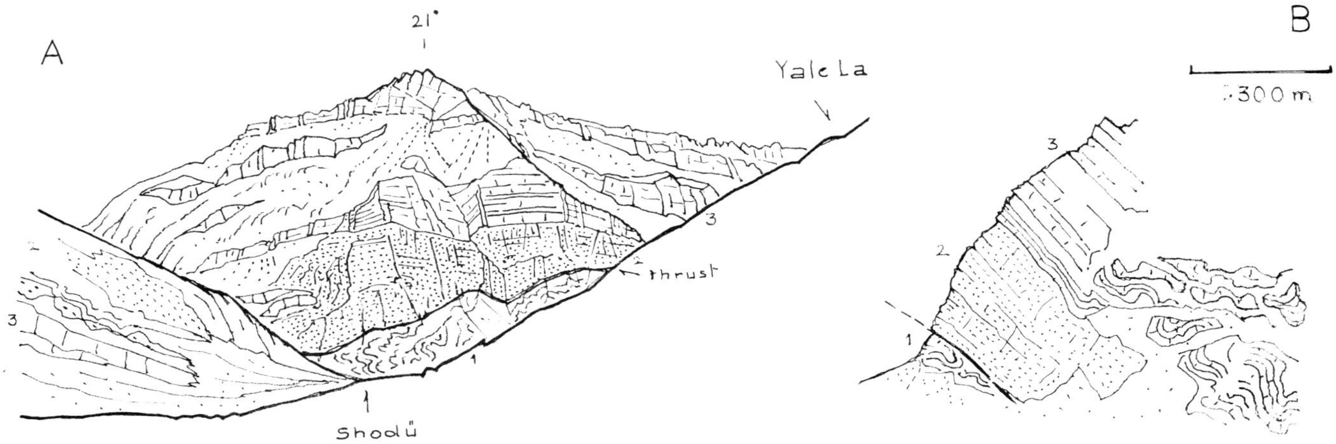

Fig. 34. The thrustzone N of Shodü, S of Yale La.
1 very irregular folded white quartzites (partly slump features),
2 shaly greywacke, with pink and white quarzite pebbles,
3 limestone, mostly dense, white grey to pink, often strongly folded and boudinaged with thin shale partings along zones of movement (traces of crinoids). Some of the folding resembles slump features (See Fig. 34. B).

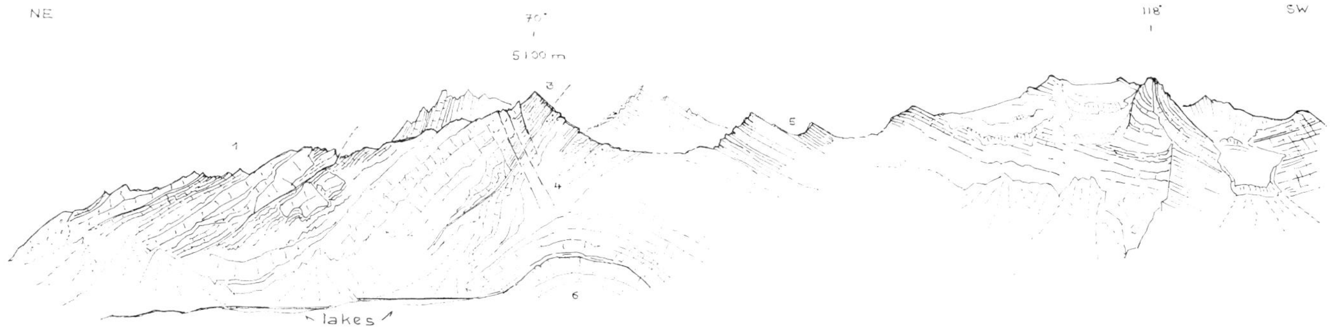

Fig. 37. The Paleozoic range SE of Chomolhari. Structural complications in Lingshi basin.
1 calc-schists with large boudinaged limestones and some crinoids,
2 dense thick bedded limestones with ammonites,
3 dark grey to black silty shales,
4 pebbly greywacke in silty dark grey shales and quartzites,
5 thick zone of bedded limestones with calcareous slates and calc-schists,
6 thick bedded light grey limestones, form core of eastwards plunging anticline.

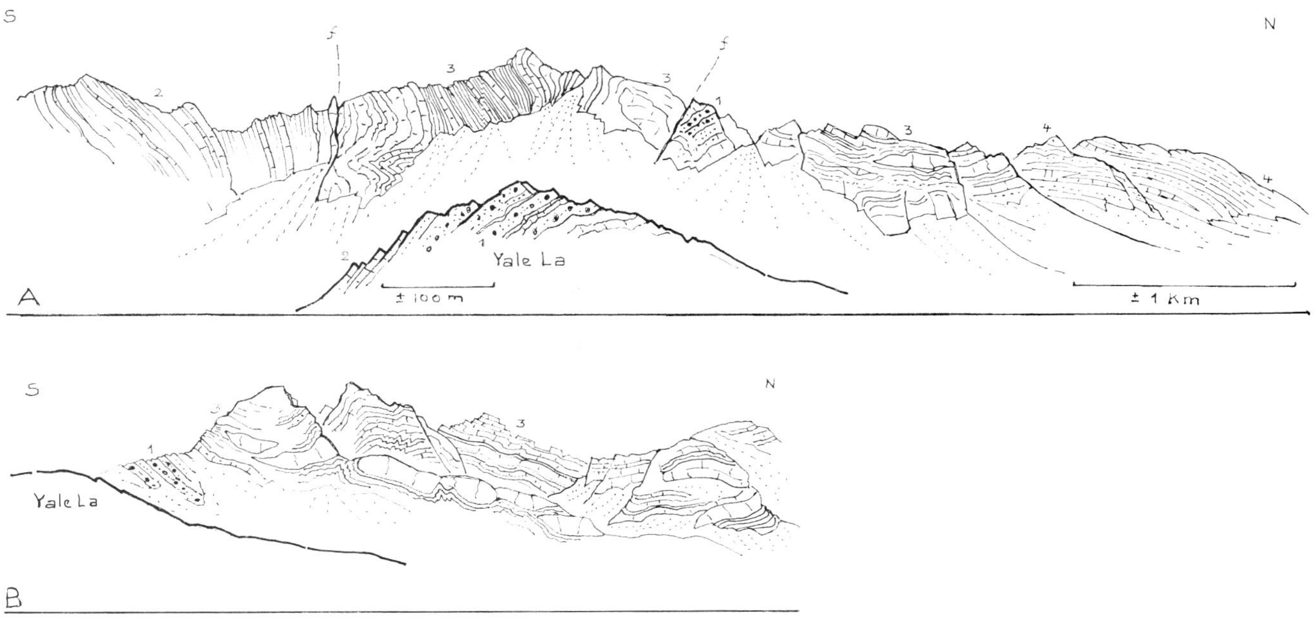

Fig. 36. Geological section of Yale La antiform. *A* view towards W and *B* projected eastern part of pass.

1 conglomerate and coarse quartzite with thin black silt partings: pebbles 2-5cm, subrounded, mostly dark grey quartz, siltstones, quartzites, arcoses, all in quartzitic matrix, with some carbonate (brown) cement, and locally sandy lumachelles;

2 calcareous slates and well bedded limestones;

3 dark silty shales and calc-schists with irregular limestones, often as large boudins, with intercalated thinbedded fine dense limestones and greywacke quartzites. Locally crinoidal limestone lenses;

4 northwards increasing black silts and calc-schists with some quartzites belonging already to Mesozoic Lingshi basin.

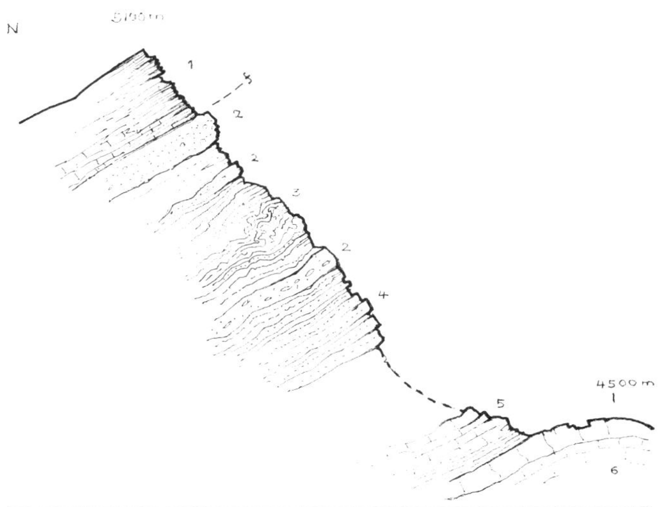

Fig. 38. Detailed composite section of peak 5100 m (SE Chomolhari) (fig. 37.).

1 Black silty shales with some limestone bands increasing northwards and partly cut out by local fault,

2 olive grey pebbly greywacke (2–5 cm quartz pebbles),

3 strongly subfolded phyllites with thin quartzite bands,

4 well bedded quartzites,

5 limestones and calc-schists,

6 thick bedded light grey limestones.

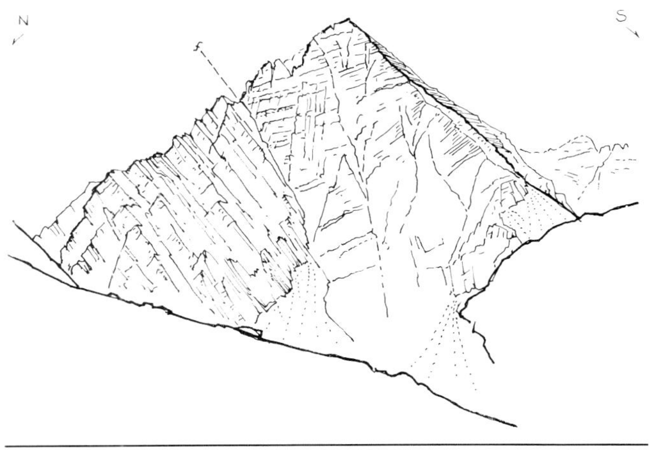

Fig. 39. Kunshupake mountain S of Chomolhari range. Faulted well bedded lower Mesozoic carbonate and shale formations, representing SW end of Lingshi basin.

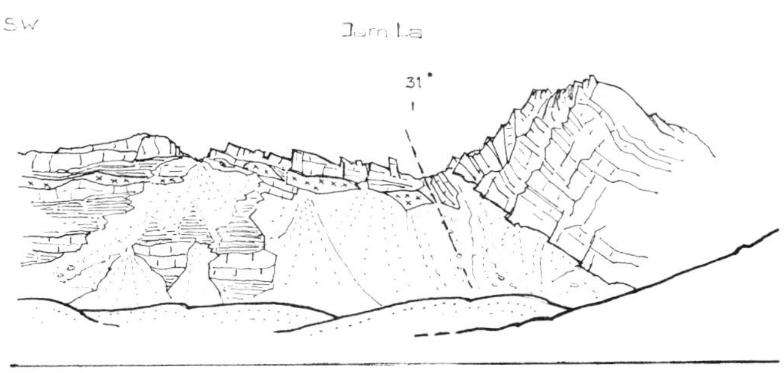

the marbles have moved outwards. We thus observe Mesozoic sediments downfaulted against the older limestones.

A more complete Tethyan sequence is preserved on the S side of the Lingshi basin above the limestones of the Wang Chu which we have followed up to Shödu (fig. 27, 128). From here we continue the section northwards over the Yale La into the centre of the Lingshi basin. The calc-schists and marbles of the Wang Chu gorge are overlain by thin-bedded pinkish quartzites and gray shales, the latter have undergone intense disharmonic folding. The axial planes dip northwards and the fold axis plunge to the E. This deformation coincides with a major thrust pushing

Fig. 40. Thick bedded marmorized limestones alternating with black silty shales and intruded by lenticular tourmaline granite layers. The sediments strike over Jam La into Tibet. N of Lingshi.

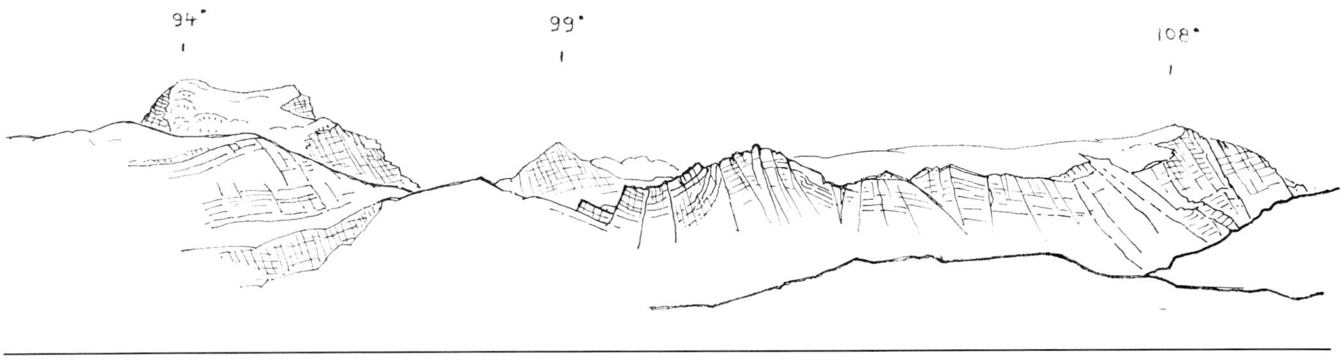

Fig. 41. The eastwards rising Lingshi basin in the Kang Bum massif, seen from Pt. 5110, S of Chomolhari. Gently folded Mesozoic sediments.

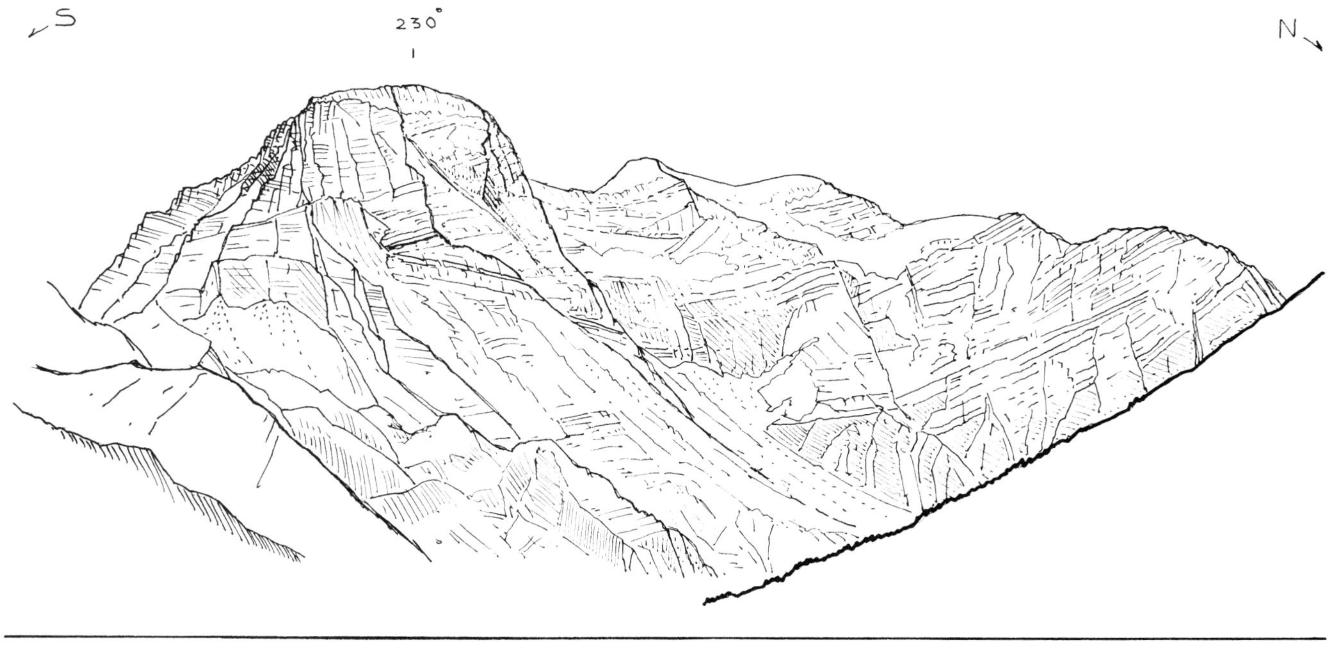

Fig. 42. The synclinal form of Kang Bum, seen from Gaza. The upper part is formed by the lower sedimentary series of the eastwards rising Lingshi basin. Possible are isoclinal folds and local imbrications.

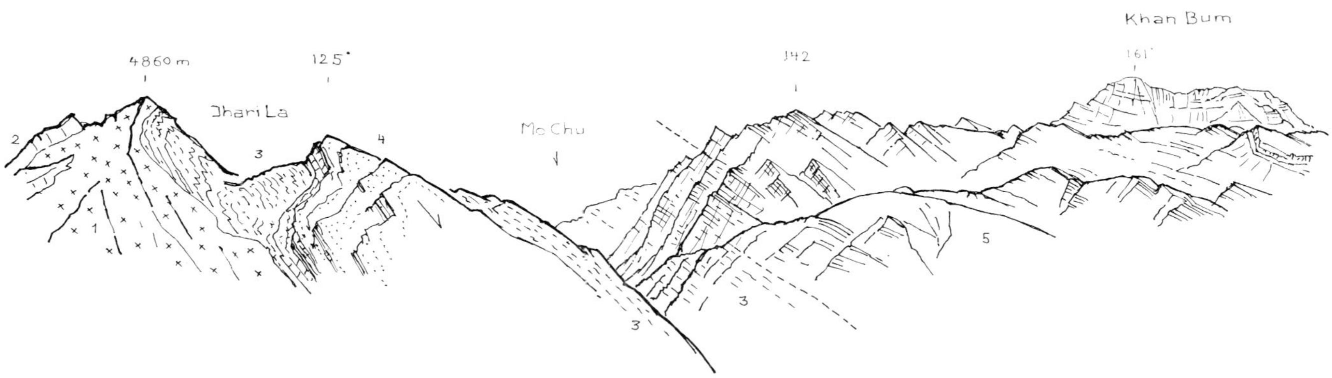

Fig. 43. The eastwards rise of the Lingshi basin, seen from W of Jhari La. *1* leucogranite, *2* marbles, *3* biotite-psammite gneiss, *4* quartzite, *5* Lingshi basin sediments.

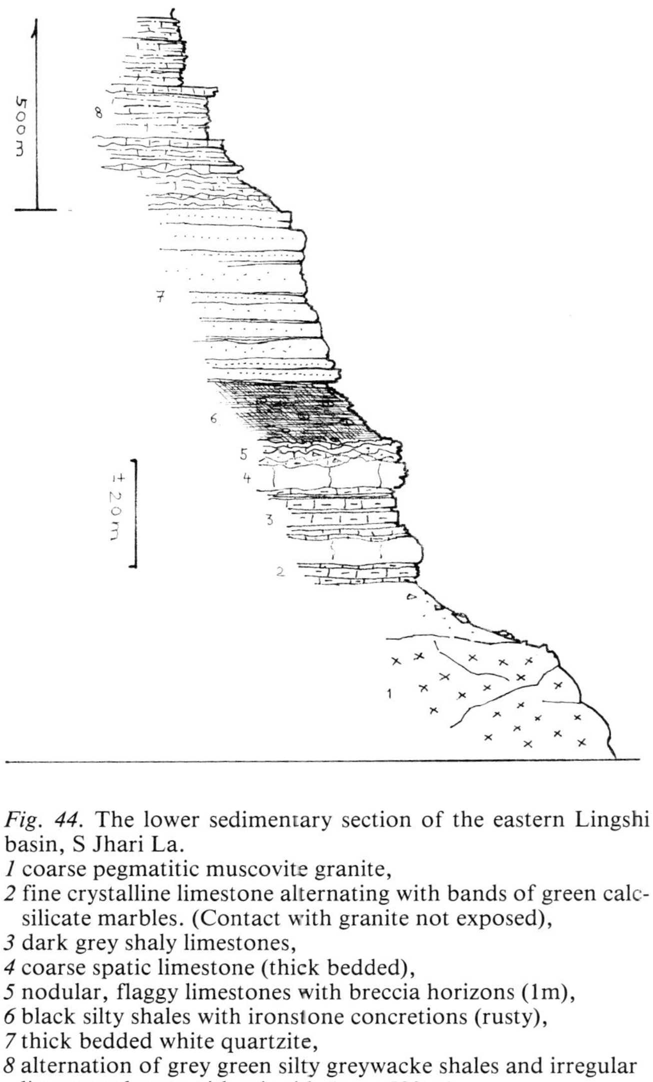

Fig. 44. The lower sedimentary section of the eastern Lingshi basin, S Jhari La.
1 coarse pegmatitic muscovite granite,
2 fine crystalline limestone alternating with bands of green calc-silicate marbles. (Contact with granite not exposed),
3 dark grey shaly limestones,
4 coarse spatic limestone (thick bedded),
5 nodular, flaggy limestones with breccia horizons (1m),
6 black silty shales with ironstone concretions (rusty),
7 thick bedded white quartzite,
8 alternation of grey green silty greywacke shales and irregular limestone lenses with crinoids (over 500 m).

the next sequence consisting of graywackes over the carbonates (fig. 34 A + B). The tectonic contact is well exposed just north of the small community of Shödu. The graywacke beds begin here with a pebbly horizon notable for the sparce, unsorted and irregularly distributed subrounded pebbles of quartzites and rusty coloured dolomites, due to a coating of limonite from altered pyrites. They are set in a dark greenish gray matrix of unsorted roundish quartzgrains and sericite mush mixed with fine carbonaceous matter derived from bituminous argillaceous deposits. The whole aspect is not unlike some tillites, and based on our present knowledge we may actually have a tillite horizon, the age of which is doubtful (*Bhatia*, 1975). The possibility of late Precambrian or Carboniferous tillities is a problem similar to the age assignement of the Diuri boulder slates, discussed under the Lower Himalaya. However, the Shödu rocks and the Diuri boulder slates are lithologically different. The Shödu greywacke section is in sharp contact with the overlying limestones and shales.*)

The first limestones are rather dense, gray-colored and well bedded. Along the strike they can abruptly become highly contorted, reminiscent of large slumping. At the same time they are boundinaged. Apart from these complications, competent folds exist with a larger amplitude and here we note an impressive cleavage, masking to some extent the original bedding (fig. 35). In the higher part of the section the limestones are somewhat coarser grained and the first fossil traces appear, mostly as small undeterminable lumachelles and some crinoides. Within the limestone calcareous slates increase in abundance and the regional N dip is gradually steepening up to a vertical fault. From here northwards we enter the complicated antiform of the Yale La (fig. 36). On the pass one finds again conglomerate horizons differing somewhat from the lower graywacke types by the content of subangular black siliceous limestones and quartzitic siltstones. Intercalated are black phyllites and brown quartzites together with thin irregular limestones containing badly preserved lumachelles. The higher beds are conspicuous for their thick irregular limestones

*) From this area T.M. *Ganesan* has reported a Rajmahal flora (Gondwana). Private communication to J.B. *Auden* (1981)

Phot. 9. The S face of Chomolhari main peak (7320 m). Well bedded marbles, schists and gneisses intruded by leucogranites. Foreground ruin of old Dzong of unknown origin.

which are often tectonically transformed into giant boudins. These are interbedded with grey and red calc-schists. Some of the limestones are dense and are covered with very irregularly distributed argillaceous films. Graywackes and dark grey phyllites are still present between the limestones, but decrease in abundance from S to N. Upwards in the sequence calc-schists and more regularly bedded limestones contain again some poorly preserved crinoids. They are possibly Palaeozoic forms of the Camerata type. The calcareous sequence is followed by more argillaceous horizons, increasing to thick bodies of black silty slates. They still alternate with limestones. In one of the black, slithly micaceous slates *Hänny* discovered the well-preserved negative of the Norian ammonite Parajuvavites.

The higher beds are mostly well-bedded to platy silty shales and marls with fine calcareous siltstones, most of it with olive-grey weathering and originally dark-grey to black. They are situated in the deepest part of the Lingshi basin and form rounded hills subject to an extensive solifluction which masks the primary contacts of the various horizons. Some fine sandy to silty micaceous horizons contain fragments of Perisphinctes and lamellibranchs resembling Pseudomonotis and Carditas, without allowing generic determinations. They may represent Jurassic horizons, since the higher silts and shales include faunas of Cretaceous affinities. NW of Lingshi Dzong layers of fossiliferous black spathic limestones occur in black shales. They contain several lamellibranchs such as Trigonias and am-

Phot. 10. The S wall of the eastern Chomolhari Range. Alternation of leucogranites (white), marbles (gray), biotite fels, schists and gneisses (dark gray). The wall is over 2000 m high.

monites, of which the Hoplitaceae and Hoplitidae suggest an Albian age. Another Cretaceous fauna was found in the platy siltstones of the Nakchu La, with ammonites probably representing Probysteroceras (Goodhallites?) and other small compressed forms, possibly Mortoniceras or Hyseroceras also of Albian age. I am indebted to *R. Truempy* and *B. Ziegler* (Zürich) for a preliminary determination of this fossiliferous material.

The Yale La antiform of mostly Paleozoic rocks forms a conspicuous range of higher limestone mountains with peaks between 5000 and 6000 m (fig. 37). Just S of Chomolhari outcrops again the quartzitic conglomerates of Yale La together with highly contorted black shales and quartzitic sandstones. They are overlain by the typical giant limestone boudins and followed by calc-schists leading to the Mesozoic section (fig. 38, 39) (phot. 13).

The presence of Middle Cretaceous in the shales of the Lingshi basin, for which I proposed the name of Lingshi formation, is of great interest. They seem to represent a different facies from the Tibetan Kampa Group and are more akin with the widespread, somewhat Flysch-like Jurassic-Triassic of parts of southern Tibet. The Lingshi basin is certainly an area where detailed stratigraphic studies are warranted. No younger rocks were noted during our traverses. It is not known if a direct connection of the Lingshi basin with the sediments of the Phari-Tuna region did exist. This was only noted in the older, metamorphic sediments. A connection seemed only possible through the depressions between the Chomolhari range and

Phot. 11. Flow folds in Chomolhari marble with folded and streched quartz veins. Concordant intrusion of leucogranite.

the high mountains of the Masang Kang region in the NE (fig. 40) and the Tremo La area in the SW.

It is surprising how the S face of the high ranges in NE Sikkim, in the head waters of the Sebo Chu, are an exact replica of the lithology seen in the S face of the Chomolhari range. *Auden* (1935) observed banded paragneisses with calc-silicates traversed by sills, some dykes and smaller stocks of fine grained tourmaline granite, giving a strikingly banded appearance to the rock slopes. This area is separated from the Chomolhari equivalent by the Tuna-Phari basin, a southwards trending anomaly in connection with the Chumbi valley depression.

The saucer shaped Lingshi basin rises eastwards above the enclosing crystallines (fig. 41) (phot. 14, 15). The easternmost rim is dominated by the synform of the impressive Kang Bum massive (fig. 42). The contacts against the rising gneisses are well exposed at the Jhari La along the Lingshi-Laya track (fig. 43, 44). The Dando Gömpa carbonate facies of the lower part is strongly reduced, from over 1000 m to a few tens of meters. Also missing are the corresponding argillaceous and thick calc-schists of Chekha. The base consists here of a medium to coarse grained, pegmatitic muscovite granite of the type intrusive into the Takhtsang crystalline and well

Phot. 12. Locally discordant intrusion of leucogranite into Chomolhari marble. Near contact boudinaged siliceous nodules. Chomolhari S wall.

exposed in the Masang Kang area. It is overlain by an alternation of thick and thin bedded, greenish to white coloured fine crystalline calcareous marbles. Unfortunately the direct contact is not exposed, but judging from scree material it seems to consist of calc-silicates not unlike the contacts of the Toma La sediments (see below). This banded section is followed by dark grey, calcite veined and slaty coated limestones and coarse, spathic limestone layers reminiscent of altered crinoidal rocks. After flaggy, nodulous, grey and white coloured limestones we note a conspicuous 30 m thick intercalation of black silty and graphitic shales, topped with a very sharp contact by over 100 m of white, fine to medium grained quartzites. They grade into an over 500 m thick section of dark grey, micaceous and fine grained graywacke with irregular calcite veins and fossiliferous limestone bands. They contain some shell remnants, crinoids and sponge neddles as well as badly preserved primitive arenaceous foraminifers.

Unfortunately, no age assignement is possible with this poor fauna. Together with the lithology, a Paleozoic age is most likely. Except for the very base, the metamorphism of this section is surprisingly low. With this sedimentary sequence the Lingshi basin ends towards the E.

Phot. 13. Well bedded sediments of the western Lingshi basin (Kunshupake range). Locally faulted (right side of picture).

Phot. 14. Well bedded sediments of the eastwards rising Lingshi basin cut by the Lingshi Chu, tributary of the Mo Chu (under clouds). View to SE. (Kang Bum Mt. to the right) Appr. limit of well bedded gneisses (below) and Lingshi sediments (above)...

Phot. 15. The Kang Bum Mt. with the eastwards rising Lingshi basin sediments in foreground. View to S.

14

15

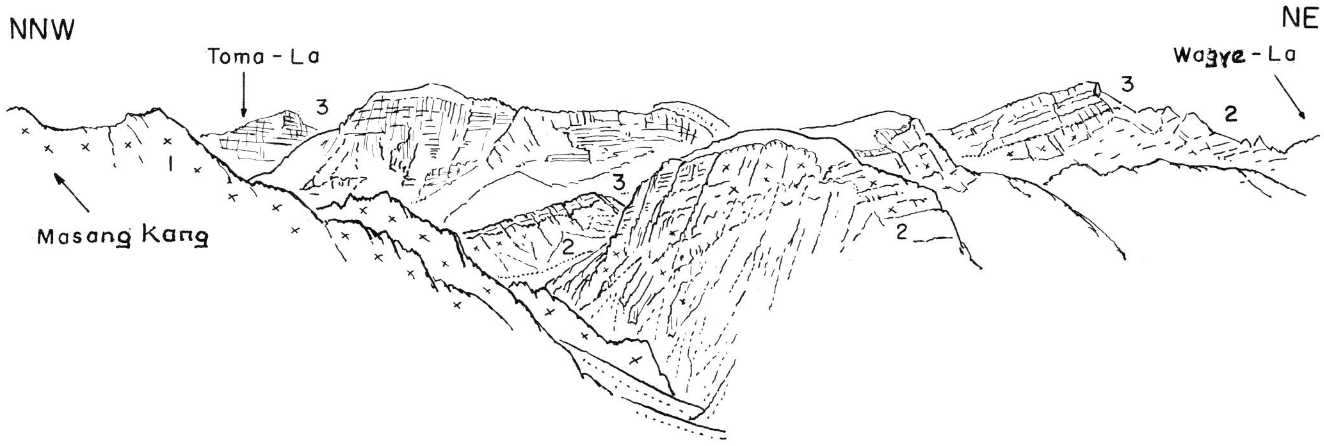

Fig. 45. The Toma La sedimentary cover on the Masang Kang migmatitic gneisses. *1* granite and gneisses of Masang Kang, *2* Psammite gneisses and migmatitic gneisses, *3* Toma La calc-schists and limestones.

E.2. The Toma La, Lunana and northeastern sedimentary belts

I found the next section of Tethyan sediments about 40 km to the ENE at Toma La (fig. 45), N of the impressive Masang Kang mountain, and again on top of the wild Tibetan border range on the N side of Lunana in northern-most Bhutan. These latter sediments were only seen on top of the 7000 m cliffs and some samples were also collected from the scree on the glaciers below (fig. 50) (phot. 16). The Toma La, the N Lunana and the northeastern sections strictly represent the southern edge of the wide Kampa Dzong basin of Tibet and do not belong to a separate and more localized basin as in Lingshi.

In the Toma La area the lower part of the Tethyan sediments is very well exposed up to the Tibetan border, forming a peculiar arid landscape, partly covered by large glaciers and their moraines (fig. 46). Here the basal contact of the sedimentary pile is visible. The widespread migmatized biotite gneisses of the Masang Kang region dip gently northwards (fig. 45) and are conformably overlain by the Toma La sediments. Towards the contact the migmatitic and partly granitized biotite gneisses change upwards to biotite psammite gneisses, still alternating with the migmatites. Upon this thick bedded sequence transgress irregular calc-schists to schistose and nodulous calc-silicates. They contain a wild mixture of larger diopside crystals, cordierite grains, bands of biotite and mid-basic plagioclases in a groundmass of irregular calcite grains, most of them conspicuously twisted and bent. It appears that the calc-schists contains reworked calc-silicate material and is itself apparently again metamorphosed, as indicated by the biotite bands. Strangely

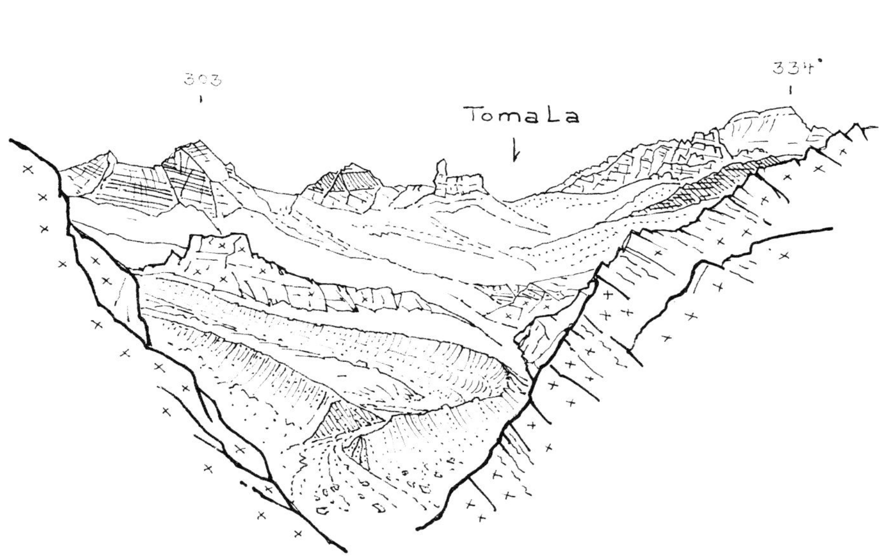

Fig. 46. Toma La seen from SE. Migmatitic gneisses covered by Paleozoic sediments forming the watershed. Note recent outbreak of glacial lake.

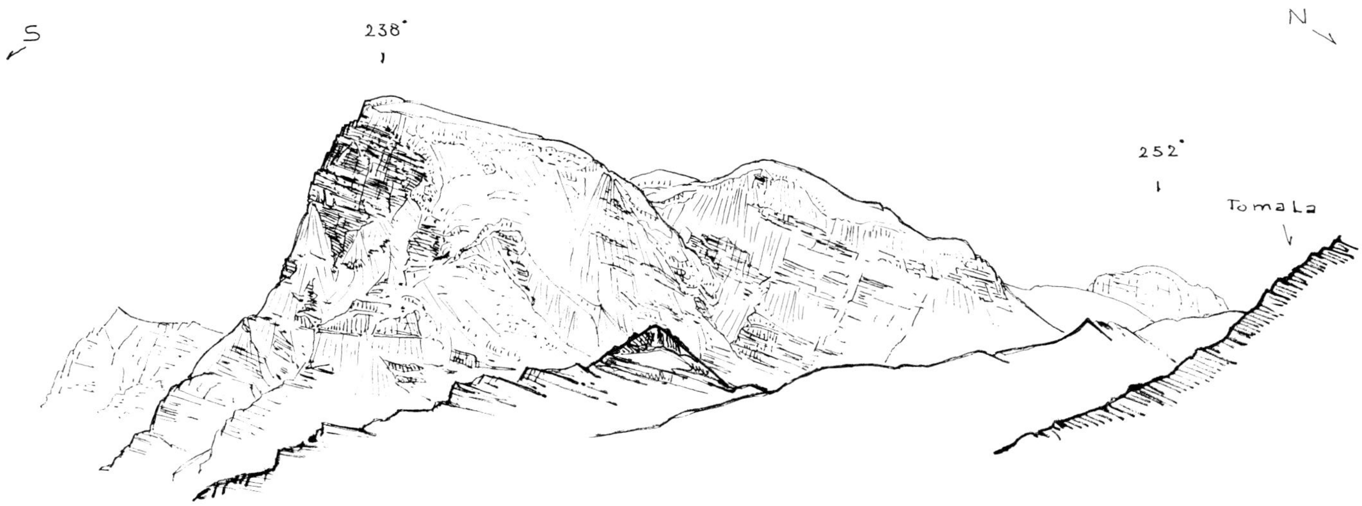

Fig. 47. The SE flank of Masang Kang. Northwards dipping migmatitic gneisses with biotite-psammite gneiss intercalations. Forms basal plate of Tibetan Tethys sediments of Toma La.

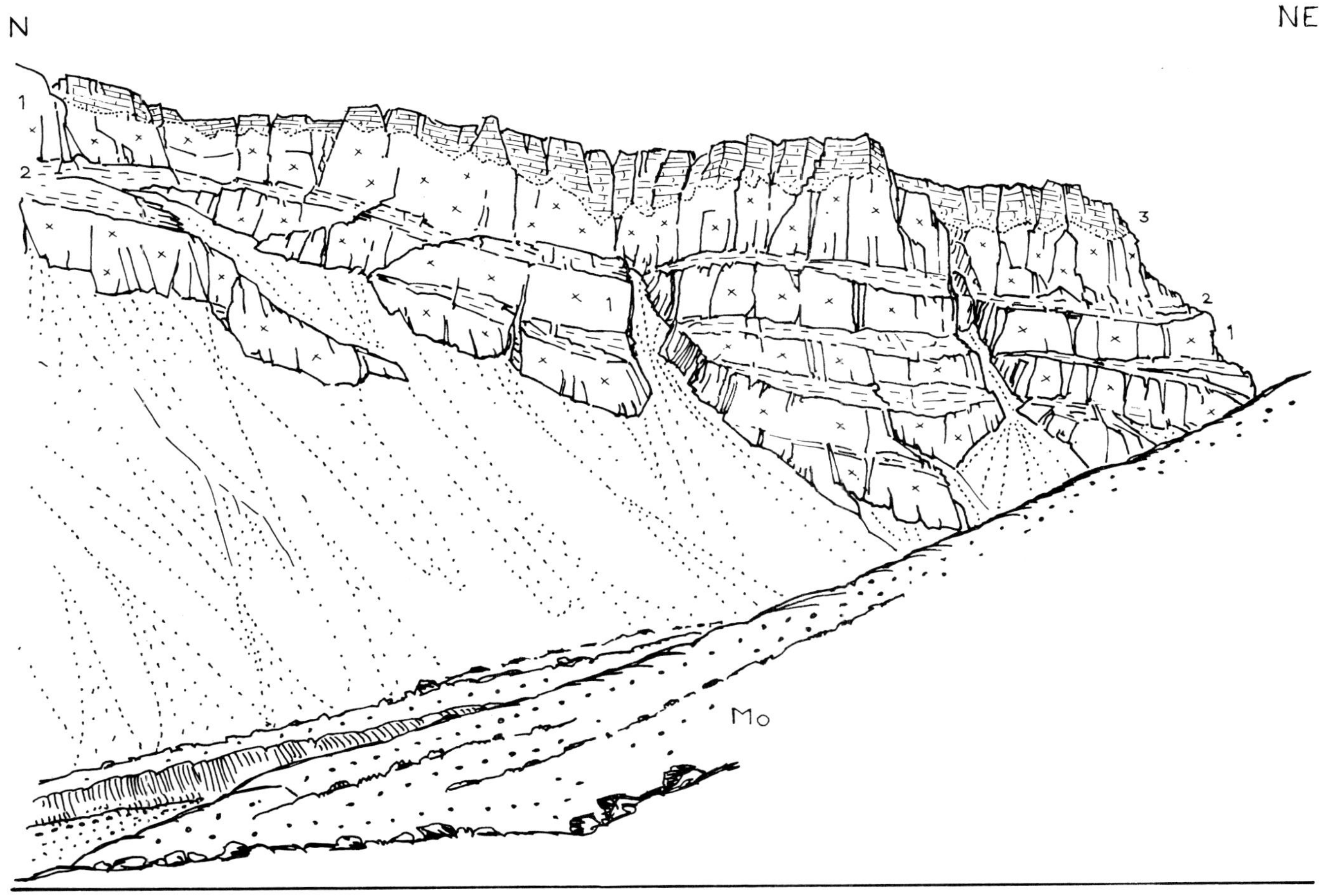

Fig. 48. The transgression of the Toma La sediments on the gneisses of Masang Kang. *1* migmatitic gneisses, *2* biotite psammite gneisses and schists, *3* basal Toma La sediments (met. calc-schists), *Mo* = moraines of the Thanza stage.

enough, real migmatitic relics from the underlying crystalline rocks seem to be missing, and the relation of the sediments with the gneisses is still somewhat obscure. The regional contact is conformable, but in detail the surface of the crystalline rocks is irregular and the calc-schists and calc-silicates seem to fill some pockets (fig. 48). Irregular patches of white calcite together with this conspicuous mixture suggest a metamorphosed caliche, which would however demand an exposed surface of the crystalline, which is difficult to explain. No younger dykes have been noted to intrude the sedimentary section of Toma La.

The basal calc-schists are followed by well bedded yellow marbles with some calc-silicates, overlain by 20 m of muscovite bearing quartzite (fig. 49). Above this and with a sharp contact follows a black biotite graywacke with larger biotite phenocrysts, without however the sieve like inclusions typical for the Budhi type schists (Chekha area). The flat lying graywacke is sliced by a marked vertical cleavage. With a rather irregular contact begin carbonates still with quartz, some plagioclases and similar biotite porphyroblasts. They grade into dark and white banded, fine grained marbles and calc-arenites. The latter were quite fossiliferous but the recrystallization allows only vague determinations such as ostracodes, molluscs or brachiopod shells, echinoderm remnants and questionable primitive arenaceous foraminifers. Some peculiar forms resemble «Globosites and Nubecularites» mentioned by Russian authors from the older Paleozoic. I am particularly thankful to J.P. *Beckmann* (Zürich) for the patience in looking at these badly preserved fossil remnants. The calc-arenites are topped by a most conspicuous band of irregularly arranged dolomitic clay pellets within a dark claystone matrix. This horizon is bordered by a normal alternation of light and dark grey coloured dolomitic claystone bands. This striking intercalation is followed by well bedded dark limestones and graphitic slates, nodulous yellow and dark grey mottled calcareous and dolomitic siltstones. The latter extend up to the pass and are coverd by spatic crinoidal limestones, grey, nodulous fossiliferous limestones and thickbedded dolomitic limestones forming the peculiar rock pinnacle characteristic of the Toma La. The same formation continues into Tibet. These top limestones are rich in crinoids and larger shell fragments, some probably from brachipods. Apart from some questionable algal remnants J.P. *Beckmann* recognized a stretched fragment of a nautiloid. Tentatively a Silurian to Devonian age could be assigned to the Toma La limestones. The metamorphism producing biotites ends upwards with the marble layers. The sediments on the pass show no metamorphism at all (phot. 17).

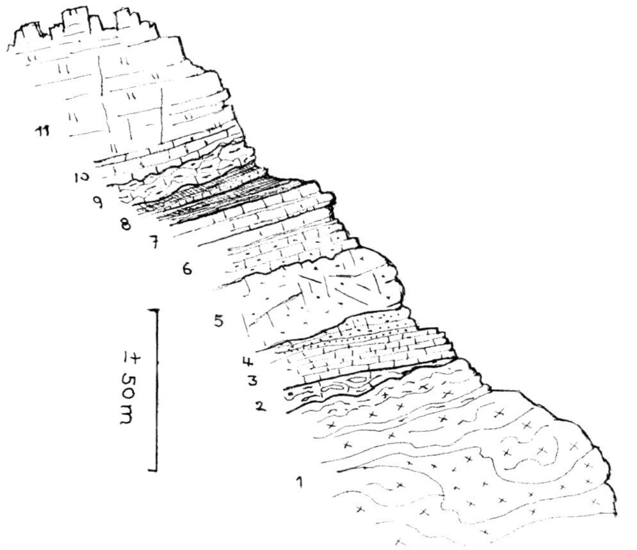

Fig. 49. Composite sedimentary section on top of migmatitic gneisses in Toma La area.
1 migmatitic gneisses with local granite intrusions. Towards the top more banded,
2 irregular calc-schists with calc-silicate nodules, sharply transgressing the gneisses (locally missing),
3 yellow, well bedded marble,
4 muscovite bearing qurtzites,
5 fine micaceous (biotit) dark grey greywacke, (marked steep cleavage),
6 sandy, marmorized limestone with biotite,
7 well bedded grey, partly dense limestone,
8 black slates with dark limestone bands, horizones with «hyerogliphs»,
9 nodulous siltstone (calcareous),
10 light grey spatic limestones with algal and crinoid remnants,
11 light yellowish-grey dolomitic limestone.

From Toma La to the E the Tethyan sediments just top the 7000 m high border range north of Lunana. Careful field-glass observations and collected specimens from the foot of the impassable wild and glaciated barrier show a conformable contact of banded calc-silicates alternating with tourmaline granite sills on top of migmatitic gneisses of the Masang Kang type. The basal sedimentary suite with lit par lit tourmaline granite intrusions is reminiscent of the sequence of the Chomolhari S wall. Phlogopite marbles come from this lower section. They are followed by dark bands of graphitic slates and fine grained, thick bedded and grey banded marbles. Yellow weathered siliceous spots and small lenses are somewhat reminiscent of fossil remains (crinoids?). The light yellowish weathered marbles form walls covered by the thick summital ice cap (fig. 50, phot. 16).

A last remnant of Tethyan sediments from the main Tibetan basin just outcrop again at the remotest northeastern corner of Bhutan, N of the legendary Sengge Dzong region. These outcrops have been seen from the 5000 m high Gwen La, exposed at the Bodu La (fig. 51), and again through the Kuru Chu gorge in mountains already at or just beyond the Tibetan border (fig. 52). The main sediments consist of well bedded black slates of considerable thickness and wide distribution beyond the Tibetan border, particularly in the upper Kuru Chu river system E of the spetacular 7600 m high Künla Kangri range (fig. 53). They also form the northern peak of the Melacachu (Garula Kang) group (NE of Sengge Dzong) and the mountains beyond (fig. 101). The

Phot. 17. The Toma La (5350 m). Dolomitic limestones form wild pinnacles. To left the 7200 m high Masang Kang with its NE face.

black slates were collected in the river gravels of the Kuru Chu just below the impassable gorge at Takhpang. They are rich in fine grained biotite, graphitic matter, some quartz, plagioclase and magnetic ore. Apart from the rich amount of graphite they resemble the black biotite slates of the Toma La section, except that they are considerably thicker. They may extend further northwards into the Trias and even the Jurassic which in the Lingshi basin is already approaching a black shale facies.

E.3. The Tang Chu basin

In addition to the sedimentary sections directly connected with the Tethyan deposits of southern Tibet we discovered in 1963 a fossiliferous sequence in Central Bhutan, preserved in a wide synform, the Tang Chu basin. It consists of a depression underlain by the main crystalline thrust sheet, here complicated by several secondary thrusts, and would therefore represent a southern remnant of the Tethyan Himalaya with apparently a more southern facies. Structurally and partly stratigraphically it is similar to the Pulchauki-Chandragiri basin on top of the Kathmandu Nappe in Central Nepal (*Stöcklin* and *Bhattarai*, in press; *Stöcklin*, 1980: *Bordet et al.*, 1969).

Not unlike the Lingshi basin, the Tang Chu sediments follow conformably above a metamorphosed argillaceous section developing from the crystallines. The crystalline rocks consist here of two mica-sillimanite gneisses with old marble intercalations and locally intruded by young tourmaline granites. They are conformably followed by biotite garnet schists and the typical biotite porphyroblastic schists known from Chekha (Budhi schists) (fig. 54).

In the hills W of the Tang Chu valley we recognize above the biotite schists thin quartzites and some calc-schists alternating with a thick section of chlorite schists and chlorite phyllites surprisingly similar to the Daling schists. The upwards decrease of metamorphism from sillimanite to biotite and greenschist facies is particularly well displayed here. To the S, within the Tang Chu gorge, a conspicuous quartz conglomerate is intercalated in some thicker quartzite layers. The well rounded pebbles, ranging from fist to head size, consist of white quartzite in a gritty quartz rich matrix. The quartzites alternate with dark grey phyllitic slates, some already with the well-known cross-biotites. After some calc-schists occur the fossiliferous section shown in fig. 55. Black, micaceous slaty schists are rich in fenestellas, brachiopods and some trilobites while the intercalated limestones contain crinoids, and at an other locality some corals. The fauna, though compressed

Fig. 50. The base of the Tibetan sediments in the border range, eastern Lunana (seen from pt. 4350 on large moraine). The basal part consists of strongly migmatized gneisses with local granitic mobilisates. In the middle part well bedded biotite schists, biotite-psammite gneisses and tourmaline granite intrusions. The top is formed by banded marbles, calc-silicates and biotite schists, forming the base of the Paleozoic sediments. (Flattish northern dips.)

Phot. 16. The 7000 m high range of northern Lunana, topped by the basal sediments of Tibetan Tethys. Base consists of migmatites, marbles intruded by leucogranites. Corresp. to eastern Table Mt. of fig. 50.

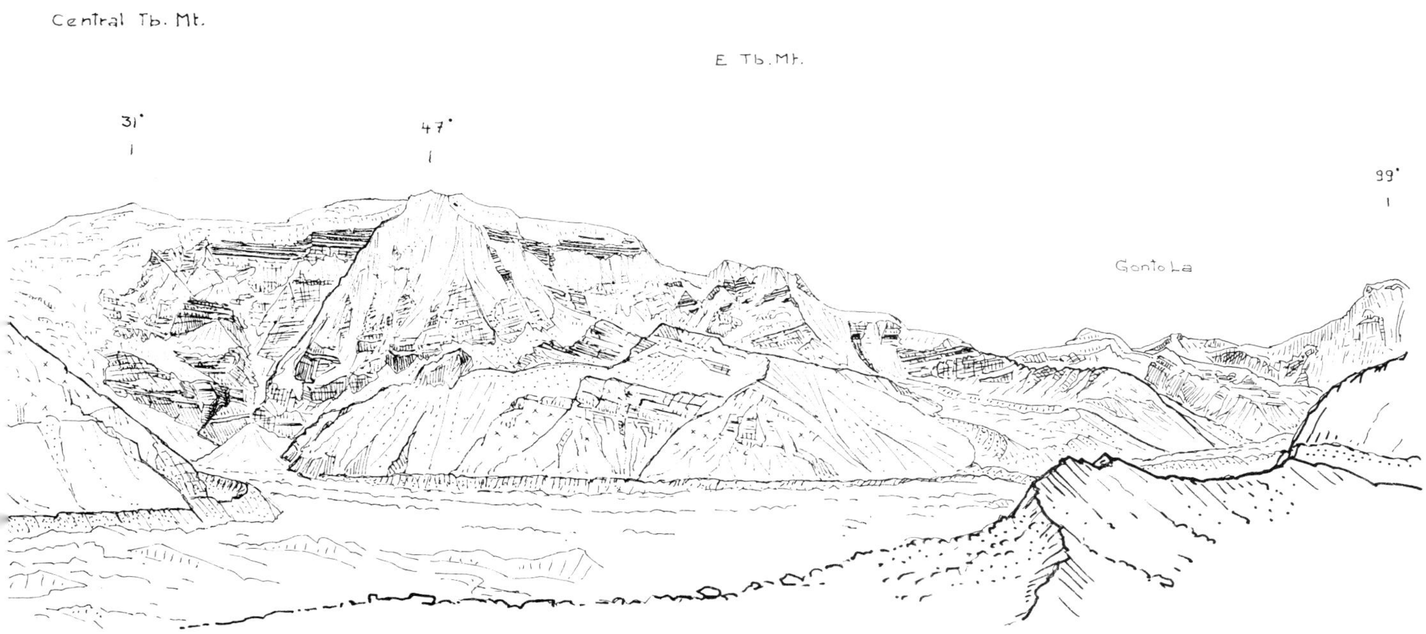
Central Tb. Mt.
E Tb. Mt.
31°
47°
99°
Gonio La

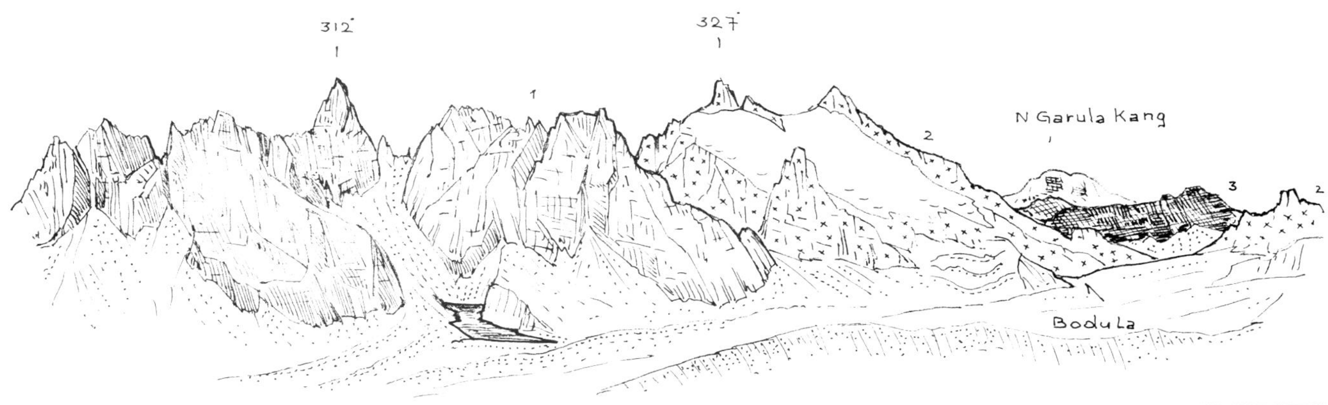

Fig. 51. The wild border range N of Gwen La. *1* vertically fractured biotite migmatite gneiss, *2* leucogranite, *3* black slates.

Fig. 52. General view of upper Kuru Chu towards Tibet. Seen from Neh. On slopes near Takpang marble zones with biotite schists and gneisses, in upper part some tourmaline granite intrusions. On right background, wild migmatic gneiss mountains between Kuru Chu and Sengge Dzong. Towards Tibet indications of black sediments (see detail Fig. A).

A. Last visible outcrops in Upper Kuru Chu in Tibet. Seen from Neh.: black well bedded formations (fine black silty micaceous slates? (samples from river). Probable base of sedimentary section.

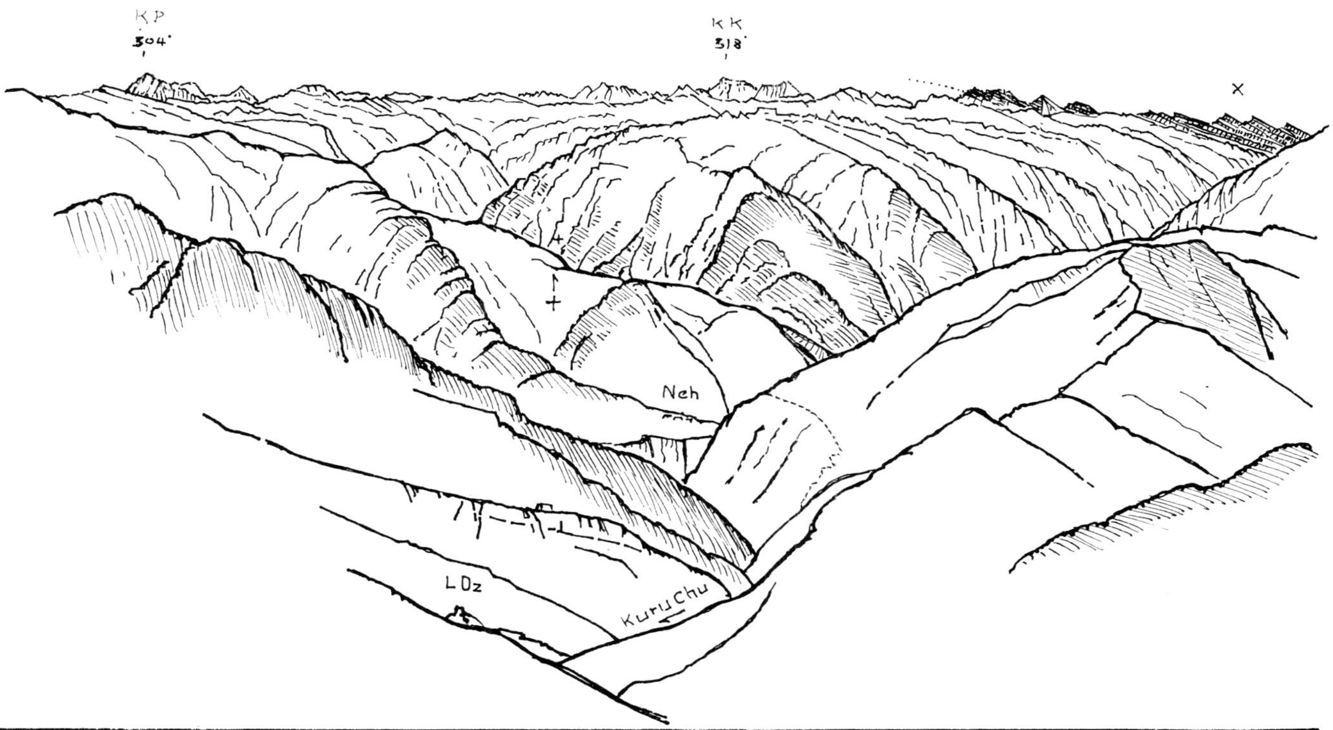

Fig. 53. General view towards NW covering the Kuru Chu valley above Lhüntse Dzong. Outstanding peneplain at 4000 m. *KP*= Kankarpünzum, *KK*= Künla Kangri, *X*= black slate zone, + = hot spring, *L DZ*= Lhüntse Dzong.

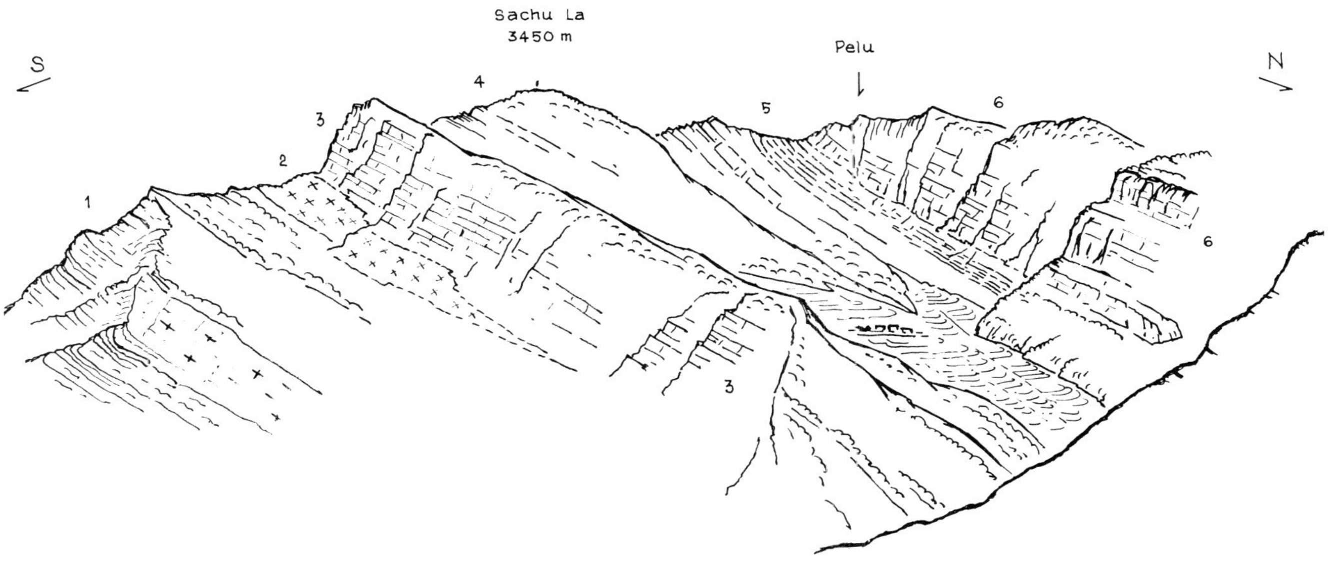

Fig. 54. The Tang Chu formation in the Pelu area (seen from the ancient roofing slate mine). *1* two mica gneisses with sillimanite schists, *2* sill like intrusion of tourmaline bearing leucogranite, *3* «old» marbles (within crystalline section), *4* phyllitic zone, *5* fossiliferous shales and limestones, *6* upper marbles forming the base of the roofing slates.

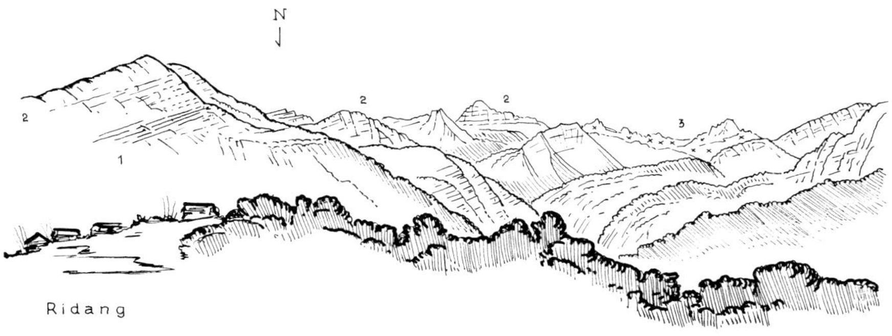

Fig. 56. The Tang Chu valley seen from Ridang.
1 horizon of roofing slates,
2 limestone marbles with some schists intercalated,
3 crystalline from the base of Tang Chu, rising northwards and intruded by leucogranites.

and in a weak greenschist metamorphic facies was studied by G. *Termier*, the corals by H. *Flügel*. Based on the brachiopod Atrypides and Elythides a Givetian/Frasnian age is suggested (Middle to Upper Devonian). The corals indicate a probable Devonian age. For more details I refer to the special publication by *Termier* and *Gansser* (1974) (see also *Singh*, 1973). Following above the fossiliferous beds we note a thick section of banded, sugary limestones with shaly intercalations and topped by most conspicuous black slates. These slates are well exposed in old quarries E of the small village of Belu. The slates are covered again by sugary limestones, derived from recrystallized originally spatic carbonate with crinoidal remnants. Dark grey spatic limestones are locally preserved and some of the calcite crystals contain crinoidal sections. These limestones can be over 400 m thick and form the higher summits (fig. 56).

The same formations run into the Tang Chu valley. The biotite porphyroblast schists are particularly well developed here (phot. 8). The olive green biotite porphyroblasts are rich in quartz inclusions and the matrix consists of fine quartz and chlorite with magnetite crystals showing well preserved crystal faces. They crop out in the lower part of the valley with thin bands of white marble, white quartzites and chlorite schists of Daling type. Gradually the thick marble horizons reappear in the upper valley from the W together with a thick belt of black slates which is covered again by marbles. The

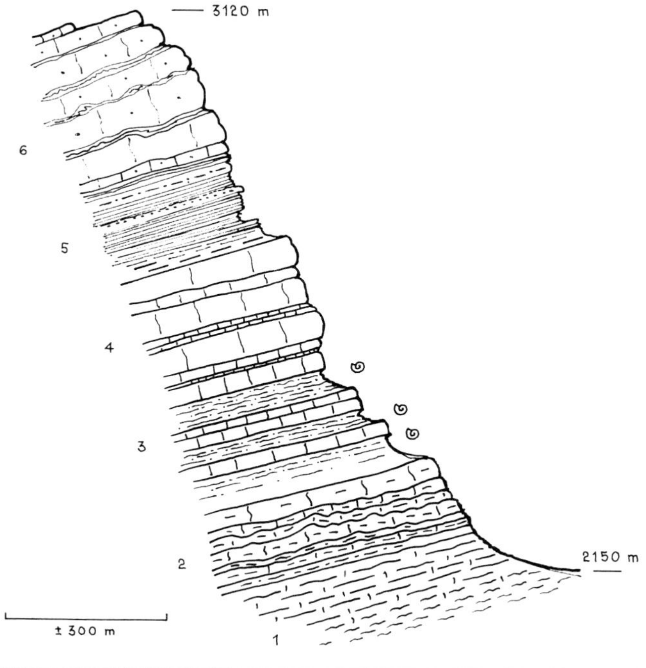

Fig. 55. The Tang Chu formation with the main fossiliferous beds.
1 phyllites and calc-schists, the base still with small biotites,
2 limestone marbles, somewhat schistose,
3 argillaceous schists with fenestellas and brachiopods alternating with crinoidal limestones,
4 saccharoidal limestone marbles,
5 black slates (roofing slate type),
6 limestone marbles alternating with schistose slates.

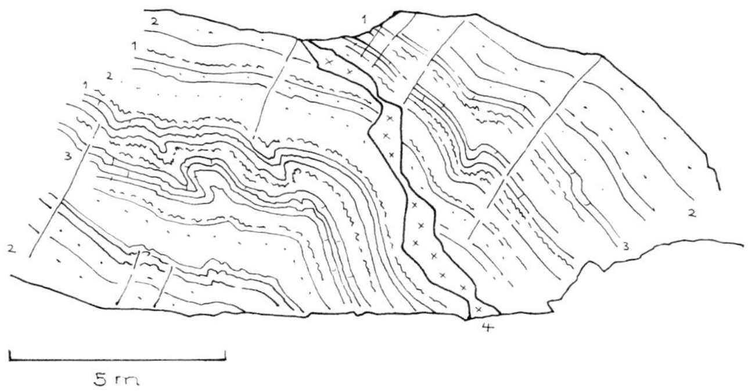

Fig. 57. Folded chlorite-biotite schists with cross biotite (*1*) banded quartzites (*2*) and thin carbonate layers (*3*) cut by pegmatites (*4*). Chumdobchi confluence in Tang Chu valley.

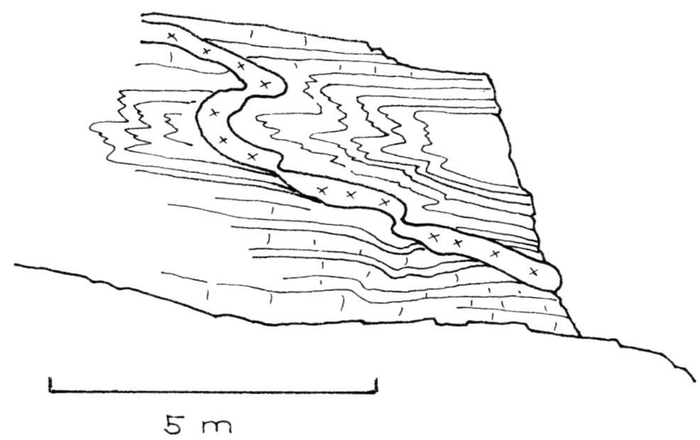

Fig. 58. Folded calc-schists and aplite-granite dykes near leucogranite intrusion, upper Tang Chu.

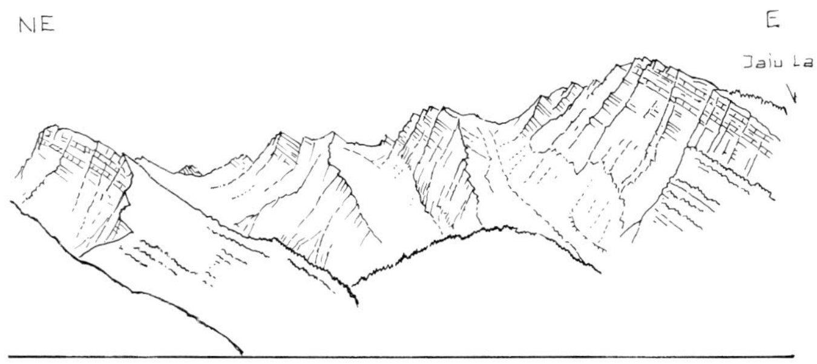

Fig. 60. The eastern flank of the upper Tang Chu valley with marbles and calc-schists. Regional SSW dip.

calcareous sections seem to have increased and form impressive walls along the steep western side of the valley. Towards the middle part of the Tang Chu valley one notes a gradual increase in metamorphism (fig. 57). Some of the marbles show flow folding and at the same time pegmatitic and aplitic granite dykes cut the folds and some are subsequently strongly boudinaged (fig. 58, 59). They are related to young tourmaline-muscovite granite intrusions cutting the lower schists and marbles which regionally rise to the N above the crystalline rocks which limit the sedimentary Tang Chu basin to the N (fig. 60). The granites outcrop in the gorges N of Takhsakha, the last village in the Tang Chu valley.

A possible southern extension of the sediments of the Tang Chu basin has been reported by *Singh* (1973) and is shown on the generalized sketch map of *Jangpangi* (1978). They occur in a syclinal position S of Taphu between the Pele La and the N-S striking high range of the Black Mountains. It is not certain if they are directly connected with the Tang Chu basin, but apparently the underlying schistose belt of the Chekha type rocks has here a wider extension than previously expected. *Singh* mentions compact green shales and thin limestones which yielded a rich fauna of brachiopods such as Spirifer, Spiriferina, Orthids ad Productides. He suggests an Upper Permian age. This age assignment seems uncertain, since *Singh* gives a Permo-Carboniferous age to the fauna of the Tang Chu basin, while *Gupta*, based on my own collection speeks of Devonian (1971).

Of special practical interest are the black slates from the Tang Chu formation mined since ancient times for the famous religeous engravings. They can be seen in many «mani-walls» (phot. 18), exposed to the rough changing climate without signes of alterations, while the best collection is preserved in the famous Dumtsi Lhakhang near Paro and in the Semtokha Dzong near Thimphu. After «rediscovering» the abandoned ancient slate quarry above Belu we have drawn attention to the Government of Bhutan to the excellent roofing slate quality of these metamorphosed rocks which can be easily cut and nailed. Now the exploitation has been resumed and already some houses and office buildings (forestry department) in Thimphu are covered by the attractive slates. I hope they will compete eventually with the ugly sheeted iron which is unfortunately replacing many of the old shingled roofs, particularly of the old Dzongs.

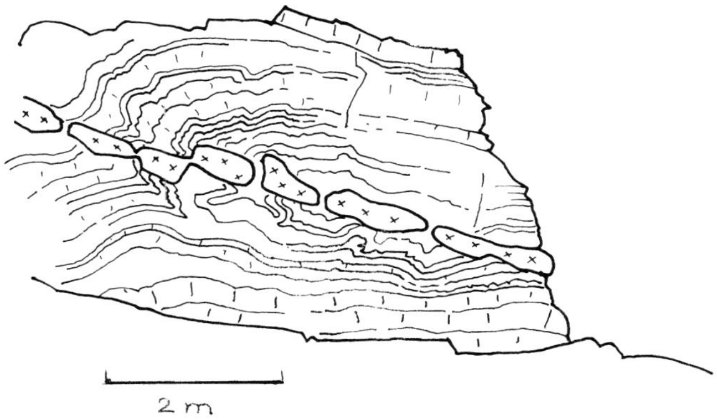

Fig. 59. Calc-schists and marbles with flow folds cut by aplite granite and subsequently boudinaged, upper Tang Chu valley.

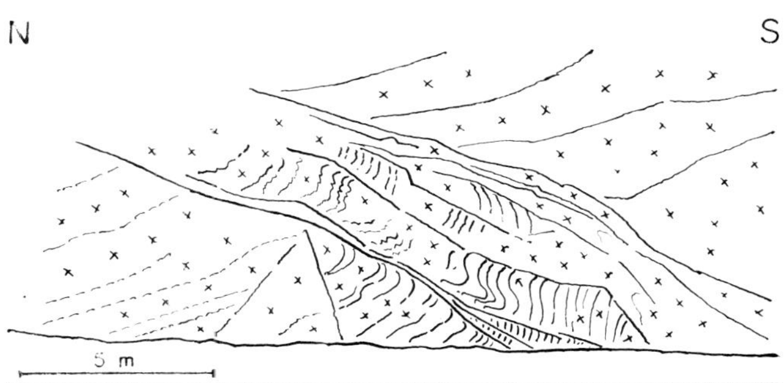

Fig. 61. Shear zones in the granit gneiss of Sure (See also phot. Nr. 23).

Phot. 18. Slate carving on Tang Chu roofing slate type. Exposed for over 100 years on outside wall of small Chörten in Bumda Gömpa (3800 m) above Takhtsang.
It represents a standing manifestation of *Byam-pa* (= *Maitreya*), his hands in the gesture of teaching. (pers. inform. by Blanche C. *Olschak*).

F. The Crystalline Units

The division between the Lower and Higher Himalaya is here not such a clear morphological separation as in other parts of the range. In Bhutan the highest mountains coincide mostly with the border ranges towards Tibet, a fact also characteristic of the eastern Himalaya in Arunachal Pradesh (NEFA). The Main Central Thrust of the Central Himalaya is poorly defined here, and similar to Sikkim, we miss the marked change in metamorphism. We noted already how the metamorphism increases upwards within the sedimentary sections of southern Bhutan and decreases again in the uppermost part of the crystalline thrust sheets.

To trace «exactly» the crystalline thrust is here at least as difficult as in other regions of the Himalaya. One of the main reasons for this is the overprint of syn-to post genetic metamorphism, which increases upwards in the underlying sections to reach nearly the high grade metamorphism in the basal part of the thrust mass.

Only rarely does one observe a subsequent mylonitization and related retrograde metamorphism indicating some later movements along the main contact. The useful amphibolitic intercalations, often observed in the Central Himalaya just below the thrustline (*Gansser*, 1964) are present in some internal thrusts but rare at the base of the main thrust. We will further realize that the crystalline thrust sheets are not uniform, but consist of various bodies with a different composition and different types of metasedimentary intercalations. While it is possible in the Central Himalaya to separate distinct crystalline nappes such as Almora from the Main Central Thrust mass, in Bhutan such units are more compressed and imbricated.

Within the sediments of the Lower Bhutan Himalaya and separated from the crystalline units we have already mentioned biotite granites and granite gneisses commonly with a porphyroblastic texture. These crystalline bodies with tectonic contacts may have been originally intrusions and are different from the crystalline sections to be discussed below. They may correlate with some of the augengneiss sheets in eastern Sikkim below the Darjeeling gneisses and even with the peculiar Ulleri gneisses of Central Nepal below the Main Central Thrust (*Le Fort* and *Pêcher*, 1974).

I have already alluded to the large extent of the crystalline thrust masses which reach in Bhutan far to the S and cover the greater part of the territory. They are divided by the Tang Chu sedimentary basin and the Black Mountains into a western and eastern sector, the latter subdivided again by the spectacular Kuru Chu-Shumar spur. Apart from the metasedimentary cover grading into the nonmetamorphic Tethyan formations, the crystallines contain large masses, lenses and layers of metasediments represented as higher grade schists, quartzites and frequently marbles with their associated calc-silicates. Similar metasediments are characteristic for the Main Central Thrust sections in the Central Himalaya where structural conditions are less complicated and where the sedimentary cover suggest a Precambrian age (Kali river section along western Nepal border, *Heim* and *Gansser*, 1939).

In the following I will first discuss the southern frontal crystalline thrust masses from three localities: N of Phuntsholing the Chasilaka crystallines (western part), the Sure crystalline N or Geylegphu (Hatisar) (central part) and the Tashigang section (eastern part). After this I will deal with the main northern zones beginning with the wider Paro region in the W and ending in the little known northeastern Bhutan N of Tashigang.

F.1. The Chasilaka crystalline unit

We noted how along the Puntsholing section (fig. 11) the metamorphism increases and that in the highest levels it produces sheared and intensely folded muscovite biotite schists. This tectonized zone is sharply overlain by the Chasilaka granite gneiss body, and it is reasonable to place the thrust contact at the base of the first major granite gneiss layer. Though the jump in metamorphic grade is small, a marked change in the lineaments is visible. Below the thrust, the schists show fold axis striking mostly to the NW while the lineaments and small amplitude folds in the gneisses are directed to the NE in the southern part and NW in the central part, here coinciding with the intercalations of metasediments.

The Chasilaka gneisses, named after an attractive mountain village at 1950 m above sea level, cover a large tract of south-western Bhutan W of the lower Mo Chu or Sankosh river. They are most probably directly connected with the Darjeeling gneisses in the W. Along the Wang Chu valley the gneisses can be followed from S of Chasilaka for over 30 km to the S side of Chapcha. Regionally the gneiss body forms a wide basin-like synform the outline of which is well visible in the relatively high mountains E of the Wang-Chu valley. The detailed picture of this gneiss basin is more complicated. Actually two gneiss masses can be noted, a lower and southern one S of

Phot. 19. Sheared, partly migmatitic, Chasilaka gneiss. S of Chasilaka.

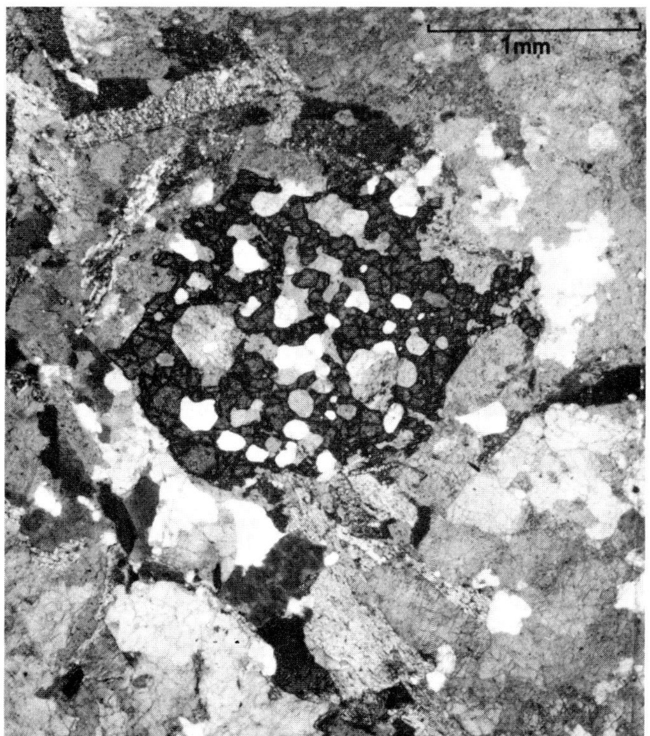

Phot. 20. Net like garnet in granulitic Chasilaka gneiss. N of Chasilaka.

Chasilaka and a northern and higher mass to the N of it (fig. 128 A). They are separated by biotite schists, lenticular quartzites and conspicuously boudinaged garnet amphibolites which normally cover the southern gneisses and dip to the NNE. In the northern gneisses occur steeply ENE dipping cross features with schists, thin quartzites, white marbles and calc-silicate bands. The latter are rich in diopside, which, apart from larger calcite crystals dominates over biotite, orthoclase, plagioclase and garnet. Together with the diopsides we note a bluish green pleochroitic hornblende.

Both the southern and northern gneisses are similar in composition. They change from well-banded and layered, often lenticular and sheared gneisses (phot. 19) into diffuse, migmatitic zones with intense sub-folding, marked by aplitic veins, which with increasing mobilization assume a more granitic aspect. Orthoclase is dominant, the plagioclase is of intermediate composition and shows some myrmekitic reaction with the orthoclase. In the northern gneisses muscovite occurs with brown biotite, kyanite and fibrolitic sillimanite related to biotite, while the southern gneisses have mostly biotite of a slightly more olive colour. Some biotites show an incipient chloritization. The quartz forms irregular lobate grains intruding all other minerals and seems to be partly introduced through mobilization. Garnet, often rich in poikiloblastic inclusions of feldspar and quartz, is always present in varying quantity and forms a most characteristic constituent, emphasizing the similarity with the Darjeeling gneisses. Fine-grained garnet-biotite granulites were observed in the northern gneisses (phot. 20).

From Phuntsholing to the W, the Chasilaka gneisses are continuing through the lower Amo Chu N of Samchi, and the Jaldhaka river into the Darjeeling gneisses. The difficulties in tracing the thrust line are here similar to those in the Phuntsholing section. *Sengupta* and *Raina* (1978) discuss this contact but include the crystalline wedges (augengneisses) in the main gneiss body. This is partly based on the correlation with the Timphu rocks, a rather unfortunate type locality which was proposed for the crystalline main units by *Nautiyal et al.* (1964) (see later).

The northern contact of the Chasilaka gneisses S of the Chapcha marble zone is not well exposed in detail, but regionally the northern Chasilaka gneisses rise over the metasedimentary sections of Chapcha. The latter are well exposed along the steep cliffs at the Chapcha spur, and lead to the metamorphic sediments of the Paro belt.

F.2. The Sure crystalline unit

From the Mo Chu river to the E the Chasilaka gneisses are strongly reduced in outcrop. In the region of Chirang N-S faulting and an abnormal NW strike direction break the normal continuation with the next, eastern gneiss mass, the Sure crystalline. *Jangpangi* (1978, p.230) mentions pyroxenites along the Sarbhang Khola which he relates to a deep seated fracture zone.

We have already discussed the Lower Himalayan sediments N of the Hatisar (Geylegphu) embayment and noted here too the gradual increase in metamorphism towards these gneisses (fig. 15). They form the hills of Sure and we called them Sure (or Suraya) gneisses (*Gansser*, 1964). Southwards they are underlain by the steeply northwards dipping biotite schists. Similar schists, with boudinaged quartzitic lenses are intercalated and again overlie the gneisses towards the N.

The main rock is a slightly migmatic muscovite-biotite-granite gneiss with some garnets, not unlike the Chasilaka gneisses. Vaguely outlined aplitic and pegmatitic dykes are rather frequent (phot. 21). Coarse muscovite-garnet-pegmatite to pegmatitic granites often include pockets of biotite-kyanite schists or intrude the schists in a most complex manner (phot. 22). The overlying biotite schists are surprisingly rich in kyanites, either associated with vein quartz, or in pure aggregates of crystals up to 10 cm long. Further northwards the cover schists become gradually less metamorphosed but are still highly stressed and sheared. The same shearing effect is visible in the Sure gneisses which are locally intensely folded and coupled with mylonitic shear zones (phot. 23, fig. 58). In the Sure region we witness again the extreme regional tectonization, indicating a S to SW directed steep thrust movement. These highly affected zones, characteristic along the southern Bhutan Lower Himalaya, contrast with the wider and gentler basin or monocline-type tectonic features following further to the N.

The Sure metasediments show a gradual decrease of metamorphism up to the biotite porphyroblast stage, similar but not quite as low grade as the Chekha formation at the base of the Tethyan sediments and the Tang Chu sedimentary basin. They are well exposed on the Sangsing La, a pass now bypassed by the new Tongsa road, and continue into the middle Tongsa Chu valley where they form a flat basin rising again to the N. Being a separate basin with sediments reminiscent of the base of the Tang Chu Paleozoic, I like to call them locally Sangsing La formation. They are most likely of a Prepaleozoic age. Its connection with the base of the small Paleozoic outcrop in the northern Black Mountains south of Taphu (*Singh*, 1973) and eventually with the Tang Chu basin seems still unknown. The Sangsing La sedimentary belt follows N of the Sure gneisses which, as we have seen, most likely correspond to the Chasilaka gneisses of western Bhutan. It is more varied and considerably less metamorphosed than the Paro metasedimentary belt. Moreover it does not underlie the gneisses, but forms their northwards dipping cover. The series starts with garnet-biotite schists, kyanite schists and garnet-sericite schists. Continuing northwards they become more phyllitic, and over garnet-bearing sericite schists one reaches a thick zone of phyllites and quartzites (Sangsing La). With the decrease of metamorphism the tectonization decreases somewhat from intense boudinage of the harder layers to folding with well developed S-directed drag folds. These are well exposed in the steep valleys of the hill zone separating the Sure region from the large Tongsa Chu valley to the N. Thicker intercalations of grey-greenish fine-grained quartzites form more compact N-dipping layers with a steep S-dipping fracture cleavage, while phyllitic intercalations show strong disharmonic folds, still with

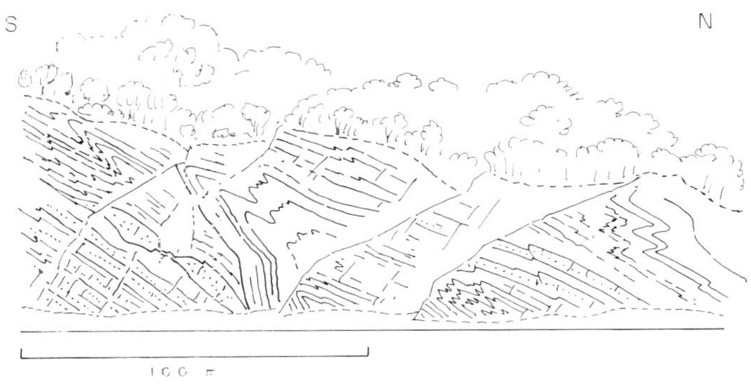

Fig. 62. Southwards vergent dragfolds in quartzites and phyllites, N of Sangsing La.

Phot. 21 Banded Sure gneiss cut by diffuse aplitic granite dyke. Sure crystalline central part.

a rather consistent southvergence. Fold axes dip mostly gently towards the W (fig. 62).

The regional metamorphism has now decreased to the phyllitic stage, but most of the sericitic phyllites show biotite porphyroblasts of a rather late formation, not unlike the biotite in the Chekha sediments. Within conspicuously banded sediments with quartzites and greenish argillaceous layers of about 5 cm thickness the biotite porphyroblasts increase in the more argillaceous part and become smaller and finally disappear in the more quartzitic horizons. This fact produces a somewhat reversed grading effect which clearly contrasts with normal cross-bedding features well visible in the quartzitic layers (phot. 24). In some instances the cross bedding is unmistakably clear and indicates that the sediments are in a normal position (phot. 25). The certainty that we deal with a normal sequence is of particular importance, in view of the fact that further north and upwards the metamorphism increases again. The biotite porphyroblasts in the argillaceous bands are rotated only slightly or not at all, thus differing from the rotated porphyroblasts in sericite schists of the Budhi type. The latter frequently occur along the Tongsa Chu valley, where thick bodies of yellowish quartzites are intercalated. There is little doubt that the green argillaceous intercalations correspond to tuffaceous horizons. Some of these bands show a strong pressure cleavage, marked by the alignment of the biotite porphyroblasts. These biotites are certainly late to postgenetic and cut some of the fine banding of the argillaceous horizons (phot. 26).

Quartzites and biotite-psammite schists form a consistently northwards dipping series free of the local disturbances met further S on both sides of Sangsing La between the Sure and Tongsa Chu valley (fig. 63). The yellowish weathered quartzites consist of translucent quartz grains with some small, biotite flakes intercalated. The bulk of the Tongsa Chu valley rocks are, however, fine, dark-coloured biotite-psammite schists with more or less argillaceous bands. In the latter the biotite can increase, forming the typical biotite porphyroblast schist. Northwards, beyond Shamgong Dzong, within a wider basin, the regional metamorphism increases slightly. Thick phyllitic horizons contain more or less well-developed biotite, garnet and some tourmaline, often as larger needles on the bedding planes.

Further to the N zones of intense local sub-

Phot. 22. Coarse muscovite-garnet pegmatites irregularly intruding kyanite bearing biotite schists. Sure crystalline, northern border.

folding coincide with larger pegmatitic and tourmaline granite dykes. Here, N of Namter, begins a thick section of northwards dipping biotite-garnet schists, psammitic gneisses and amphibolites belonging already to the front of the northern crystalline sheet of the Tongsa region (fig. 64). At the base of this section, where we noted intense folding together with cross cutting intrusions of pegmatitic tourmaline granites (fig. 65) we assume a major thrust, which, not unlike all the major tectonic contacts within the metamorphic rocks, is difficult to localize in detail. The thrust hades to the N with about 10° to 15° and would correspond to the base of the main northern crystalline unit.

F.3. The Tashigang crystalline unit

This unit overlies with a well defined thrust the Shumar-Barsong sediments (fig. 128). It is limited to the W by the northward directed sedimentary spur and cross high of the Kuru Chu-Shumar region, while the northern border is arbitrarily taken at Yangtse Dzong. Here no break exists with the main northern crystalline mass. Eastwards the Tashigang crystalline extends beyond the border into the Tawang region of Arunachal Pradesh (NEFA). Here it has been reported from the Sela Pass where migmatites and

Phot. 23. Sheared and thrusted migmatitic Sure gneiss, locally folded. Sure crystalline, central part.

two mica granites, developed from granitization, are cut by tourmaline bearing aplitic dykes. Sillimanite occurs in intercalated metapelites south of the pass. The MCT aequivalent passes about 5 km to the W of Dirang, the capital place of the Kameng district which borders Bhutan. It strikes to the NNE and dips with about 30° to the WNW (*Verma* and *Tandom*, 1976). It is thrust over the Bomdila metasediments, corresponding in a very general way to the Daling-Shumar group.

Along the Tashigang road, just N of Barsong, the southern sediments are overthrust towards the SW. With the sharp and striking change in metamorphic grade, lithology and structure as well as the intercalated wedge of the Barsong formation (p. 30), this thrust is certainly the best recognized major tectonic contact in Bhutan. It continues in a northwesterly direction along the E side of the Kuru Chu half window or spur just S of Dameche Dzong into the Kuru Chu valley S of Lhüntse Dzong. Here we noted the northern closure of the Kuru Chu spur with a marked increase in metamorphism of the underlying sediments such as marbles and quartzites resembling the Paro type metasediments (fig. 126).

S of Tashigang, the crystalline thrustmass begins with strongly sheared and laminated biotite-(muscovite) - alkalifeldspar (microcline) gneisses. Some show drag folds with irregularly mobilised quartz-feldspar veins with rims of biotite (fig. 66). The gneisses dip regionally to NNE at 20–40° and show a pronounced lineation in the same direction. Some of the larger orthoclases form augen. Biotite

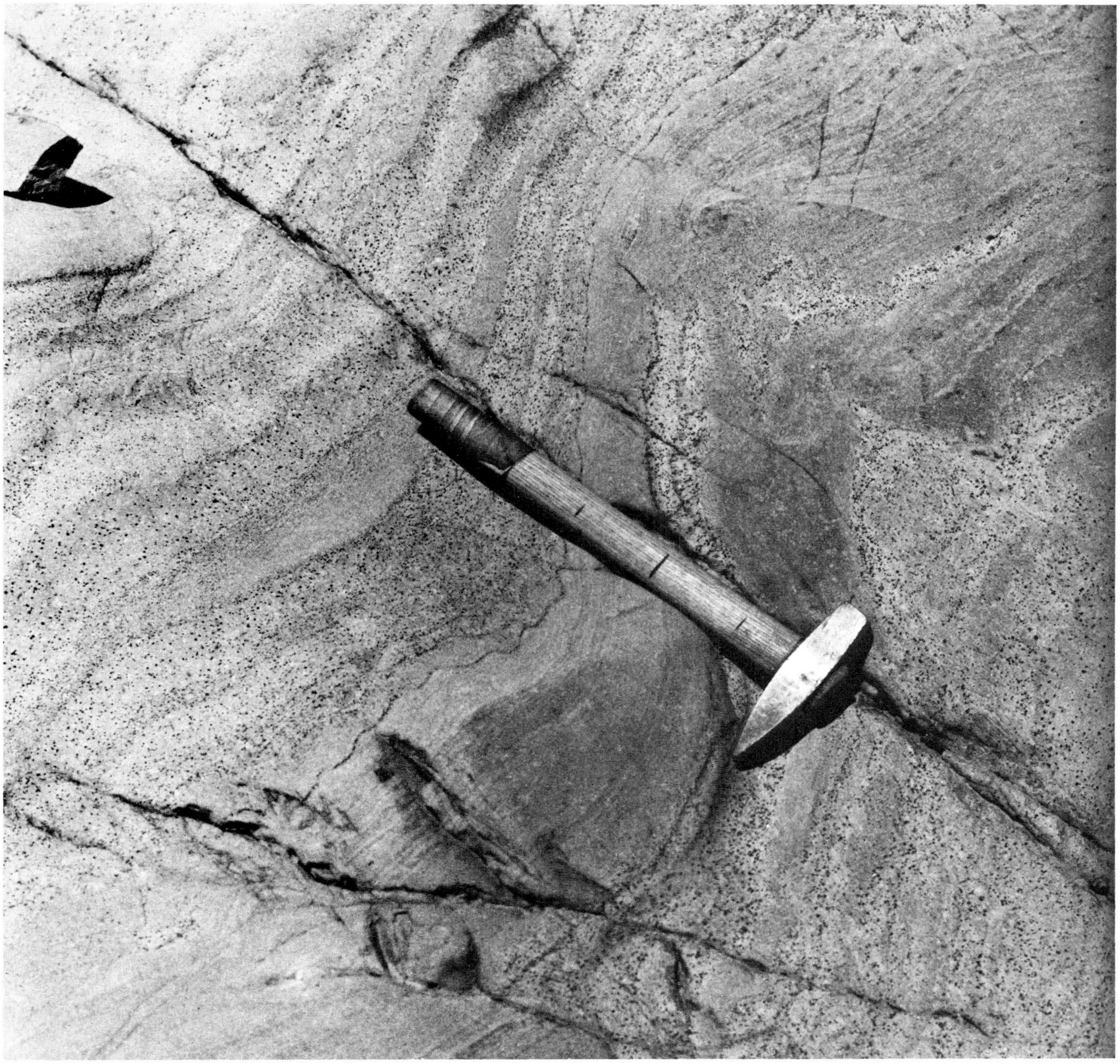

Phot. 24. Banded quartzites and green argillaceous (tuffaceous) layers. The latter show biotite porphyroblasts, decreasing towards the quartzites producing some abnormal grading effect. Note normal cut-out on upper right of picture, giving top-bottom indication. Sangsing La.

(and muscovite) underline the schistosity and the quartzes show a predominant orientation with the C-axis perpendicular to the main layering. A recrystallization under stress seems well developed along the thrust contact but no retrograde metamorphism was noted. The lineation of the biotite schists just below the thrust as well as the less metamorphosed quartzites and slates is directed to the E. Biotite gneisses continue northwards towards Tashigang still dipping to the NNE and NE while the direction of the lineation and the axis of small folds have changed towards the E. Locally some fine aplitic tourmaline granite sills intrude the gneisses. At Tashigang the gneisses become somewhat migmatitic and the metamorphic grade increases, which can be noted in the metapelites, such as sillimanite-kyanite-garnet-muscovite-biotite schists and gneisses with a varying content of plagioclase. Large kyanites are present in the neosome with quartz and some orthoclase, while the sillimanite occurs only as small fibrolitic grains attached to the biotite.

Between Tashigang and Gomkhora in the N, just below the confluence of the Yangtse Chu and the Dangme Chu, the latter coming from Tawang in the E, the crystalline rocks form a remarkable synform with a higher metamorphism, including granitization exposed in the center (fig. 128). This synform rises to the W but opens in an eastward direction S of Dangme Chu and towards the Tawang region in Arunachal Pradesh, beyond the Bhutanese border. The biotite gneisses of Tashigang dip with 50° to the N but flatten gradually towards the center

Phot. 25. Cross bedded quartzites of the Sangsing La formation. N of Sure.

Phot. 26. Folded fine quartzites and green argillaceous tuffites with biotite porphyroblasts. Sangsing La formation.

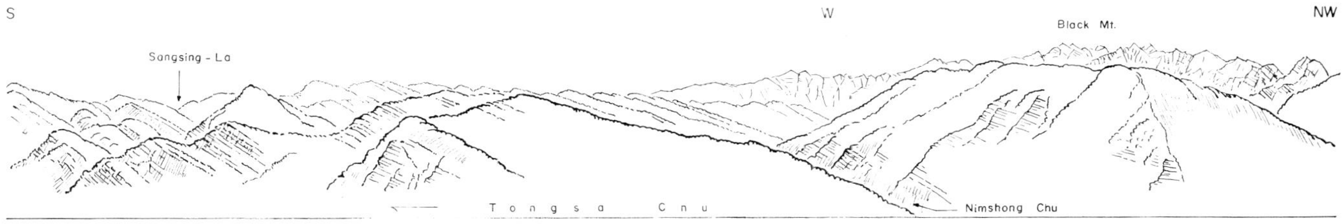

Fig. 63. Regional N dip in quartzites and psammitic schists along the Tongsa Chu. To the right the morphological culmination of the Black Moutains.

Fig. 64. Generalized section from Nada La to Namter (Mangde Chu). Note zone of increased metamorphism and dyke intrusion in middle part of section.
1 quartzites and garnet biotite phyllites (the biotites are elongated in S), *2* garnet amphibolite in garnet biotite schists, *3* biotite phyllite and sericitic quartzites, *4* psammite gneiss with granite gneiss intercalations cut by pegmatitic granites, *X* = detail in fig. 65.

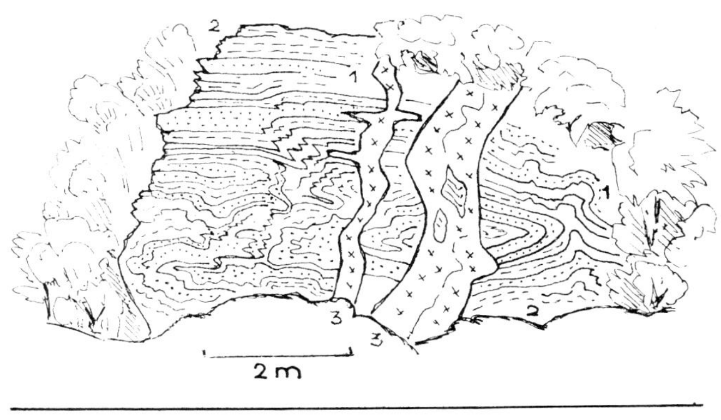

Fig. 65. Above Namter detail of folded quartzites (*1*) alternating with biotite schists (*2*) and cut by pegmatitic tourmaline granite dykes (*3*) foldaxis dipping with few degress eastwards.

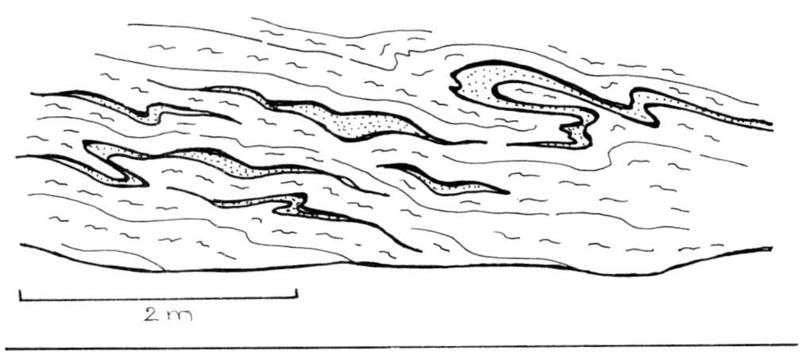

Fig. 66. Migmatitic biotite gneiss with folded, mobilized biotite-aplite granite veins. S of Yangtse Dzong.

Fig. 67. Strongly folded, striated biotite-(muscovite)-psammite gneisses. Foldaxis dipping constantly 170°/10°, axial planes eastwards. Just below confluence Yangtse Chu and Dangme Chu (Tawang Chu).

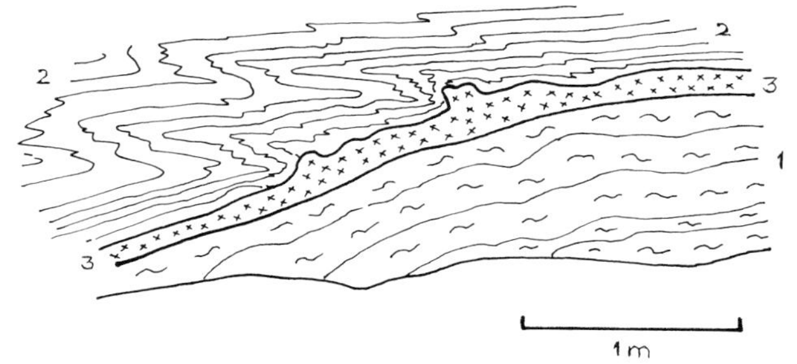

Fig. 69. Disharmonic folding between biotite-granite gneiss (*1*) and biotite schists (*2*) with thin biotite-aplite dyke intruded along contact (*3*). S of Yangtse Dzong.

of the synform. Between the migmatitic biotite-muscovite gneisses one notes layers of well bedded biotite-psammite gneisses often with thin intercalations of garnet rich aplites. In the center of the synform occur 1-2 km thick bodies of massive hornblende diorites rich in sphene and with middle basic plagioclase. They change into thick-bedded but fine grained biotite granite gneisses with aplitic veins which to the N are followed by banded augite amphibolites and fine grained granulitic diopside gneisses. The augites are fresh and the hornblendes have an olive brown pleochroism. The latter have not been derived from altered augite. The granulitic diopside gneiss is of special interest and reflects the highest metamorphic grade in this synform. It is free of hornblende with fresh diopside and plagioclase of a basic labrador type, which also forms the leucocrat part of the amphibolite. The granulitic diopside gneiss contains some rather roundish quartz grains. Both rocks are free of calcite and do not seem to be related to calc-silicate rocks.

From here to the N we enter the northern flank of the synform. We observe the same biotite (muscovite) gneisses, partly migmatitic, with veins of aplite and some coarse muscovite-biotite pegmatites, which we have mentioned from the southern part of the synform. This flank can be followed to the region of Gomkhora where quartzitic biotite-psammite gneiss dominate. They are intensely folded with axes dipping gently to the S while the axial planes hade mostly eastwards (fig. 67). Quartzes and plagioclases are strongly stretched and the biotites sometimes arranged in a polygonal way around the folds, suggesting a syn- to postgenetic recrystallization. Along this N flank of the synform the regional metamorphism has again decreased. We can follow this flank along its ENE strike into the Dangme Chu valley towards Tawang beyond the eastern border of Bhutan (fig. 68). Up to the border occur biotite gneisses, biotite-psammite gneisses and biotite-schists. Between schists and gneisses one notes some disharmonic folding and the contact can be intruded by fine grained tourmaline aplite (fig. 69).

From the confluence of the Dangme Chu and the Yangtse Chu to the N (Yangtse Chu valley) the S dipping flank continues, mostly with the same biotite gneisses already seen further S. Locally occur some intercalations of muscovite-garnet gneisses and kyanite bearing muscovite schists with quartz veins within biotite psammite gneisses. Gradually the S dip flattens and one crosses a marked domal feature where again somewhat less metamorphosed rocks outcrop in the center, such as well bedded but mostly intensely subfolded quartzitic biotite schists. The axes of the small folds trend from WSW to ENE and dip gently to the W while the megafold axes are more

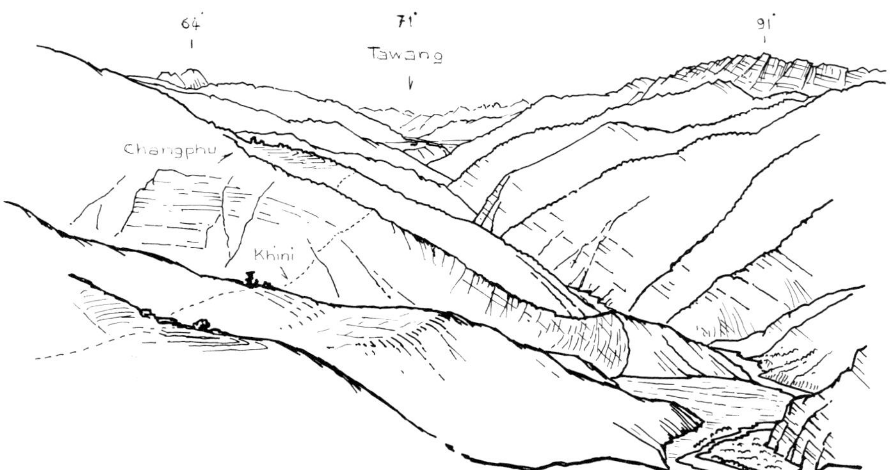

Fig. 68. The N flank of the Tashigang synform (biotite-(muscovite)-gneisses) striking along Dangme Chu into the Tawang region.

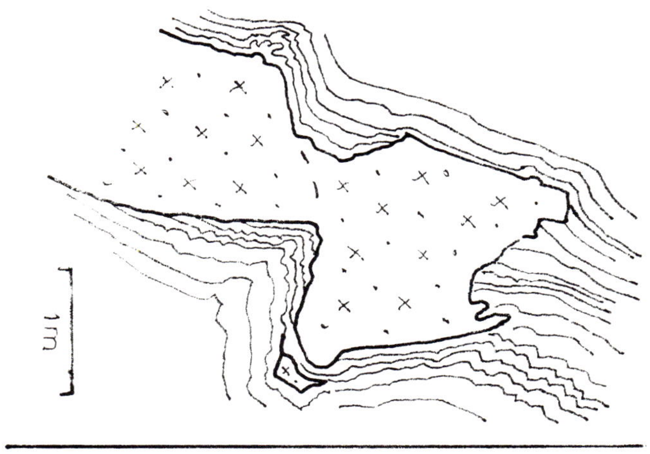

Fig. 70. Acide aplite-granite dyke cutting biotite-muscovite gneiss, locally refolded. Late genetic intrusion. S of Yangtse Dzong.

EW aligned. This window-like antiform again suggests the regional decrease in metamorphic grade downwards in this particular E Bhutan section. Biotite granite gneisses and well bedded biotite muscovite gneisses often with aplitic tourmaline granite veins (fig. 70) follow N of the antiform and continue with a marked northern dip towards Yangtse Dzong and beyond.

F.4. The Paro metasediments

High grade metamorphic metasediments are related to the crystalline units and occur in various tectonic positions within, below and also above the main crystalline. Their largest development was found in the Paro region, which has been taken as the type locality. They must be distinguished from the sediments of the Lower Himalaya as well as the basal sediments of the Tethyan sections. I cannot agree with the however interesting suggestion by *Jangpangi* (1978) that the Paro sediments be the metamorphic equivalent of the Baxa group in the Lower Himalaya. The fact that they form a «window» in the Paro region is no convincing argument, apart from the quite distinct type of sedimentary sequence. More difficult are some distinctions within the marble belts, characteristic for many intercalations within the crystalline units but also typical for the basal Tethyan sediments of the Chomolhari region (Dando Gömpa marble facies). In both sections we often find the young tourmaline granite intrusions. Since the high grade metamorphism reaches into the lower part of the Tethyan sedimentary pile and the upper limit of the crystallines is taken where the metamorphism decreases (Budhi type schists) this last distinction is rather an academic one.

F.4.1. The Paro Region

We met the Paro type metasediments at the N end of the Chasilaka crystalline unit which overrides the marble, quartzites and schists of the Chapcha region. After an EW directed antiform, the metasediments continue with a constant NNE to NE dip along the Paro Chu to Paro and the Wang Chu to Thimphu, the present capital of Bhutan. This rather gentle monocline is broken by two structural disturbances; one in the lower Paro valley striking EW, and the other in the Wang Chu above the confluence with the Paro Chu, striking NNW. Both zones expose steep dips and are characterized by boudinaged amphibolitic lenses and associated garnet bearing calc-silicate rocks with sheared biotite schists forming the main lubricant. Between the Wang Chu disturbed zone and the confluence of the Wang Chu with the Paro Chu one observes a most conspicuous granite gneiss dome, apparently a domal-folded thick laccolith consisting of laminated biotite-garnet gneiss with diffuse migmatitic schlieren. Laterally, as well as below and above this gneiss body, are garnet-staurolite-bearing muscovite-biotite schists, which constitute the main argillaceous rock type of the Paro metamorphics. Frequently the garnet schists are minutely folded and the garnets rotated. The fold axes dip southwards S of the dome and to the N on its NE side (phot. 33). Another granite gneiss body is located W of the lower Paro valley in the watershed between the Paro and Ha valleys. It consists of thickbedded muscovite-biotite augengneisses with some augen reaching up to 10 cm. Its relation to the surrounding schists and marbles is not yet known.

Most conspicuous in the Paro valley are the banded marbles, called Paro marbles. Three main marble zones can be distinguished in the Wang Chu and Paro Chu valleys. The lowest horizon crops out just N of the Chasilaka gneiss basin in the Chapcha region. It is interbedded with garnet schists and frequently contains calc-silicate layers with boudinaged quartz veins (phot. 27). The higher marbles occur in the Paro valley proper, separated by quartzites, though a tectonic repetition may actually be possible. Here the marbles are banded with white and grey layers, ranging from a few centimeters to nearly one metre (phot. 28). The banding is an original sedimentary feature marking more or less bituminous layers. The white and darker bands are coarse and fine grained respectively, in accordance with the well-known retarding effect of bitumen during recrystallization. Syngenetic folds are still visible in the banded marbles as well as primary pinch-outs of certain layers. Intercalated in the marbles are well-bedded mica schists and very coarse, pure quartzites with some small biotite flakes and large translucent quartz grains. Regionally, all the marbles and major quartzites have only gentle dips, mostly northwards or NE-directed. Disturbances are, as already indicated, sharp linear features such as steeply dipping to vertical tectonized zones.

Above the upper Paro marble at Paro Dzong

Phot. 27. Banded calc-silicates with boudinaged quartz veins in contact with garnet schists-gneisses (left side). N of Chapcha.

Phot. 28. Gray and white banded Paro marbles. Original sedimentary features are still preserved. Paro type locality.

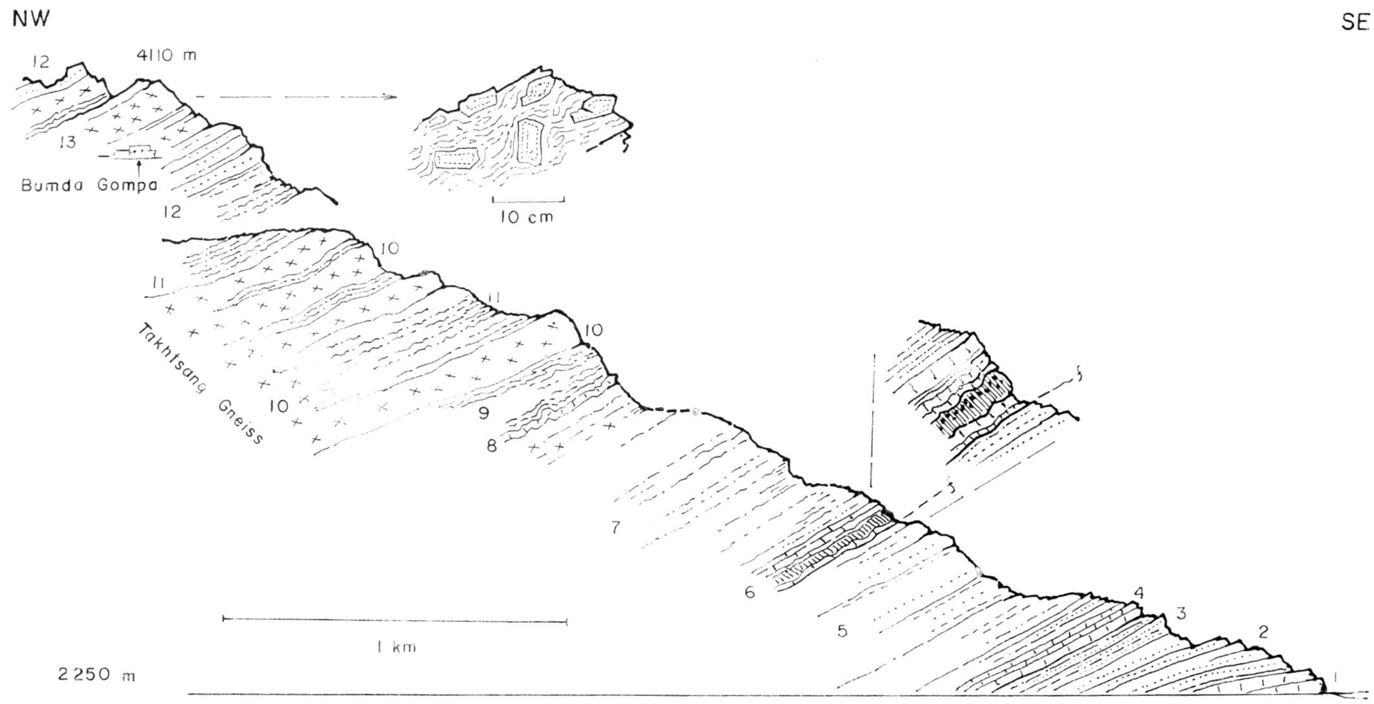

Fig. 71. Contact of Paro Zone with Takhtsang gneisses, N of Paro.
1 Paro marbles, *2* muscovite quartzites, *3* psammitic schists, *4* banded diopside marbles, *5* coarse muscovite quartzites, *6* garnet-amphibolites with calc-silicate bands (tectonic contact), *7* muscovite-biotite schists and gneisses, *8* thin calc-silicate band, *9* sillimanite gneiss, *10* biotite-muscovite-garnet gneiss with sillimanite, Takhtsang type, *11* sillimanite schists-gneisses, *12* thick bedded coarse quartzite, *13* coarse augengneiss.

Fig. 72. Generalized section through Paro type rocks between Confluence and Thimphu, Wang Chu valley.
1 garnet and staurolite bearing muscovite-biotite schists, *2* garnet rich calc-silicates, locally with garnet amphibolite lenses, *3* garnet bearing biotite-granite gneisses, locally migmatitic, *4* Paro marbles, *5* well bedded quartzites with thin micaschists, *6* scapolite bearing biotite amphibolites and actinolite schists, *7* garnet bearing muscovite-biotite-K-feldspar gneisses, *8* muscovite-biotite schists, *9* biotite psammite gneisses with garnet quartzite concretions, *10* biotite psammite gneisses with garnet bearing cordierite-sillimanite-biotite-K-feldspar gneisses, *11* Takhtsang type gneisses.

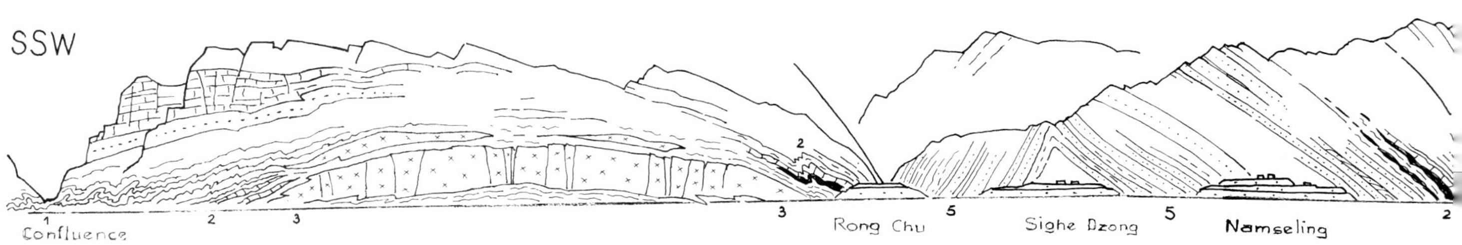

follow schistose quartzites with characteristic large silvery muscovite flakes on the bedding planes. Locally single kyanite crystals are embedded. Northwards and upwards the meso-metamorphic quartzites show a gradual increase in metamorphic grade leading towards the next higher gneiss sheet (fig. 71).

A good section of the Paro metasediments is exposed in the Wang Chu valley between Thimphu and the confluence of the Wang Chu with the Paro Chu. It points out the complexity and the tectonization of these horizons (fig. 72). We already referred to the gneiss dome and the tectonic complications but have not mentioned yet the various rock types. At the confluence outcrop intensely folded biotite-sericite bands. The fold axis dip towards SSE (phot. 29). In the lower Paro valley they are followed by a steep zone with calc-silicate bands and actinolitic amphibolite lenses, leading to a steep dislocation. This is very similar to the one in the Thimphu section to which we refer in the following, but which is not directly connected. Along the Wang Chu, below the biotite quartzite we note a southeastwards dipping section of garnet-two mica schists containing some staurolite. Intercalated are lenses of garnet rich calc-silicate. This belt forms the southward dipping cover of the granite gneiss dome already mentioned. The northern cover of the dome is similar to the southern one though the calc-silicate contains a thin sugary dolomite band together with amphibolite lenses and layers. We reach here a marked fault zone, running into the Rong Chu valley, after which follows a thick section of well bedded quartzites alternating with thin mica schists. They dip steeply to the N and probably form a sharp antiform near Sighe Dzong. The N flank, still with the same quartzites , continues up to Namseling. From here upwards the dips become steeper. Some of the quartzites show a weak cross-bedding suggesting here a reversed sequence. Gradually amphibolites are intercalated, some with large garnets. The quartzites become thinner with large silvery muscovite flakes in the schistose intercalations. They are followed by calc-silicates and amphibolites which form a steep to vertical zone with strong shearing along the contact of the various rock types. From vertical the dip turns gradually to the S still with calc-silicates, garnet-biotite amphibolites and biotite-garnet schists. Some of the biotite amphibolites are rich in scapolite, replacing the plagioclase. Leucoxene is also frequent. They are folded and faulted and border with a steep N-hading thrust against quartzites which have a peculiar type of flow folding, practically without intervening schists. The fold axes dip gently to the SE. Again due to thrusting this folded section is in contact with a northwards dipping sequence of quartzites, biotite-garnet schists, coarse quartzite and muscovite schists and some thinner zones of biotite amphibolites as well as actinolite schists. Towards Semtokha Dzong the white quartzites, often coarse grained, dominate again together with muscovite schists. Locally they contain some tourmaline. Gradually the first gneiss intercalations occur within muscovite-biotite schists and quartzites . Between Semtokha Dzong and Thimphu the gneisses increase such as larger layers of garnet bearing muscovite-biotite-alkalifeldspar gneisses. Near the main Thimphu bridge one notes small garnet bearing quartzitic concretions within biotite psammite gneisses. Well bedded biotite gneisses and garnet biotite schists with rotated garnets (phot. 30) can be followed up towards Dechenchöling. Between these well bedded biotite-psammite gneisses and garnet schists occur layers of garnet bearing cordierite-sillimanite-biotite-alkalifeldspar gneisses. These rocks are of particular interest since we realize that we have reached the high grade metamorphic sequence forming the Takhtsang crystalline unit.

We may stress the point that the contact between the Paro type sediments and the higher Takhtsang crystalline units is very gradual. All the major disturbances are within the Paro metasediments. The gradual contact passes through Thimphu where, incidentally, the rocks are not well exposed, which is one reason why I cannot agree to use Thimphu

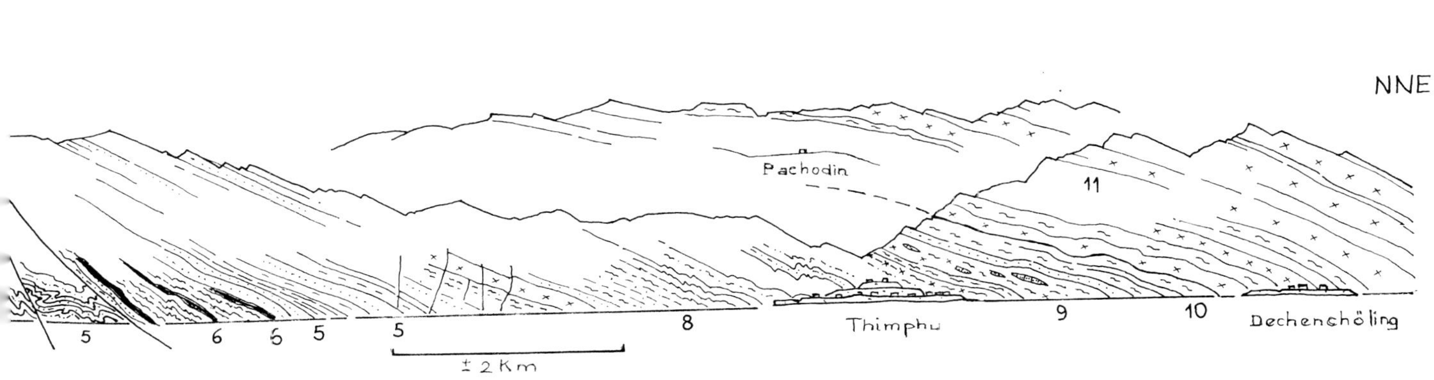

Phot. 29. Folds and lineations in banded garnet schists, locally quartzitic with biotite and sericite layers. Fold axis dip towards SSE.

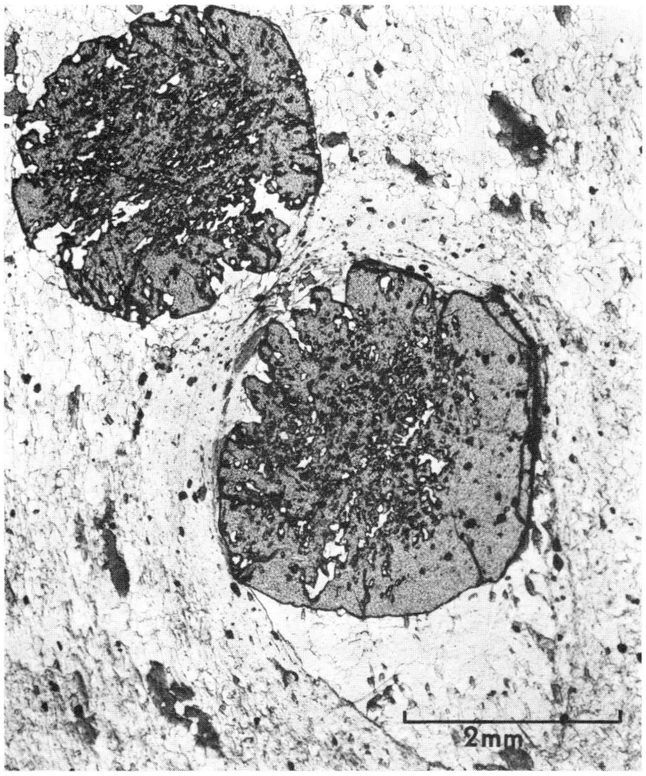

Phot. 30. Garnet biotite schists with rotated garnets compressed towards the upper left. Note quartz in pressure shadow and some corrosion on frontal part of garnet. Thimphu.

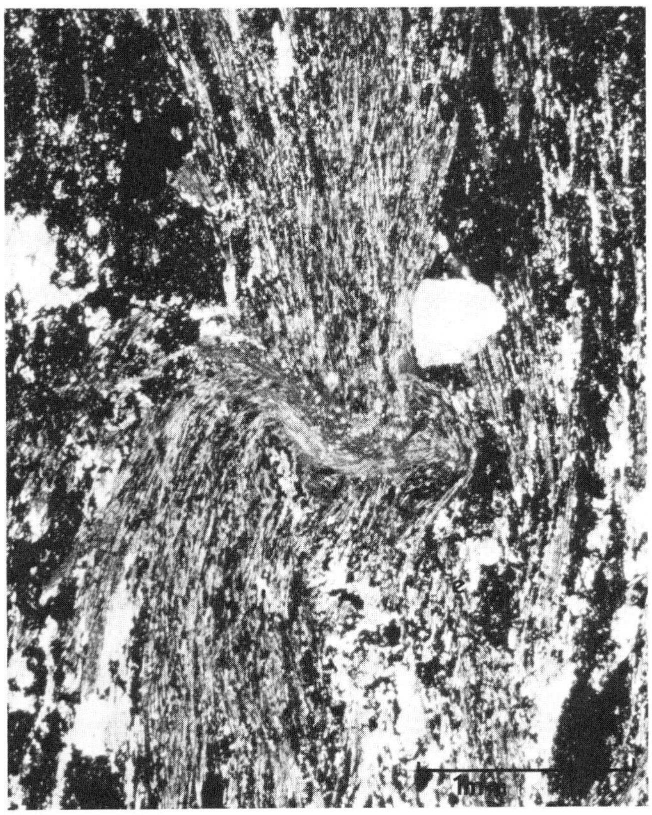

Phot. 31. Graphitic sillimanite schists, intensely folded. The sillimanite needles are sharply flexured. W Jule La, above Ha Dzong.

(inspite being the capital) as a type section for the high grade crystalline unit (*Nautiyal, et al.*, 1964), but prefer the beautiful rock manastery of Takhtsang NNW of Paro (*Gansser*, 1964) (see below). I agree to speak of a Thimphu gneiss, meaning the well exposed granite gneiss NW of Thimphu which has been quarried for the reconstruction of the majestic Tashichö Dzong, the main Bhutanese government and religious centre (see later).

The thick Paro marbles are not encountered along this section. The carbonates of the Chapcha region are far below this section, while the lower Paro marbles can be seen in the steep cliffs N of the river confluence above the Tamchö Gompa, the home of the famous first iron chain-bridge builder (Thangtong Gyalpo), which used local raw material. The marbles, corresponding to the lower Paro marbles are underlain by quartzites. They can be followed in the high slopes as far as the Rong Chu, where they are probably cut off by a crossfault.

Towards the W, the Paro metasediments continue into the Ha valley. They end about 7 km NW of the famous Ha Dzong and this contact can be followed along the watershed between the Ha Chu and the Amo Chu in the SW and S respectively. Here this border coincides with the northern end of the Chasilaka crystalline unit. The contact with the northern overlying Takhtsang gneiss unit is gradual, similar to the section N of Paro while the southern contact seems sharper, corresponding to the Chasilaka «thrust». Beyond the Paro belt on the W of the Ha valley the Takhtsang and Chasilaka type crystalline units should border each other, but the nature of this contact, also important in Sikkim, is not known. The small Paro type wedges of metasediments along the Amo Chu near Sombe Dzong (*Jangpangi*, 1978) may belong to intercalations similar to those already mentioned within the Chasilaka gneiss unit.

The Paro marbles can be followed over the Jule La into the Ha valley and are well exposed near the Ha Dzong where steep fault-zones have cut them into conspicuous steep hillocks. The marbles are white and grey banded and undistinguishable from the Paro type locality. They are underlain by muscovite quartzites locally with kyanite and alternating with some muscovite-biotite schists, psammitic gneisses and some calc-silicate bands. Along the lower slopes N of Ha Dzong and again well exposed on the pass the lower marbles contain graphitic sillimanite schists, very rich in graphite. The sillimanite forms sheaf like bundles up to 12 mm long apart from more finely distributed crystals (phot. 31). Some ghost like pale biotites are filled with graphitic matter. The graphitic schists are rather widely distributed, but since they are predestinated as shear zones, good outcrops are rare. This graphite could locally be of commercial interest if the layers were less disturbed and the localities less remote.

Towards the E the Paro metasediments disappear below the gneisses around Thimphu but crop out, at least the uppermost contact zone, in the Mo

Chu valley between Punakha in the N and Wangdüphodrang in the S. Here we note mostly micaceous quartzites, some with large, silvery muscovites similar to the base of the Takhtsang gneisses. They crop out towards Wangdüphodrang and are overlain by muscovite-biotite schists and gneisses, the latter with some sillimanite already of the Takhtsang type. Irregular and boudinaged amphibolites and calc-silicate layers indicate structural anomalies (E of Dochu La). The thick marble horizons of the Punakha region belong already to the Takhtsang zone. The whole gneiss section dips to the E and forms here the base of the overlying Tang Chu basin already discussed. Southwards the gneiss zone continues into less known regions and eventually reaches the southern gneisses in the Chirang district. The published information of *Jangpangi* (1978) of these areas is only very generalized while much more data is stored in the files of the Geological Survey of India.

F.4.2. The Cändebi metasediments

E of the Tang Chu basin and of the important Pele La outcrops within the crystalline a sequence of carbonate and pelitic metasediments. They are well exposed at the small village of Cändebi which gave the name (*Gansser*, 1964). They form a broken antiform with a marked N flank but faulted S flank. Both flanks are covered by tectonized crystalline with locally younger tourmaline granite intrusions.

The structural complications are well visible in the lowest Chusom valley on the western side of Cändebi. Towards the E, near the famous Chörten, the carbonate section is cut by a marked NS directed fault zone with phyllonitic biotite schists bordered towards the E by biotite-muscovite gneisses. The latter belong to the complicated domal antiform of the Tongsa gneiss body (see below).

In the W we note an increase in metamorphic grade along the eastern border of the overlying Tang Chu basin. E of the Pele La we still recognize the already well-known biotite porphyroblastic schists, here very rich in tourmaline. Gradually we come into biotite-psammite gneisses with some pegmatitic veins and then begin the first marble bands alternating with calc-silicate layers, both intruded by small bodies of tourmaline granite. Some of the marbles expose isoclinal flow folding. Most conspicuous are intercalations of highly graphitic biotite-garnet schists (phot. 32) wich, apart from the rotated garnets, are very similar to the graphitic schists accompanying the Paro marbles in the Ha valley. At Cändebi we note a northwards dipping zone of fine to medium grained sugary dolomitic marbles which form horizons up to 300 m thick. They alternate with finely banded white and yellow marbles, whose banding is partly accentuated by thin dolomitic seams. Intercalated are biotite and garnet phyllites. Compared to the Paro marbles the Cändebi section is more dolomitic.

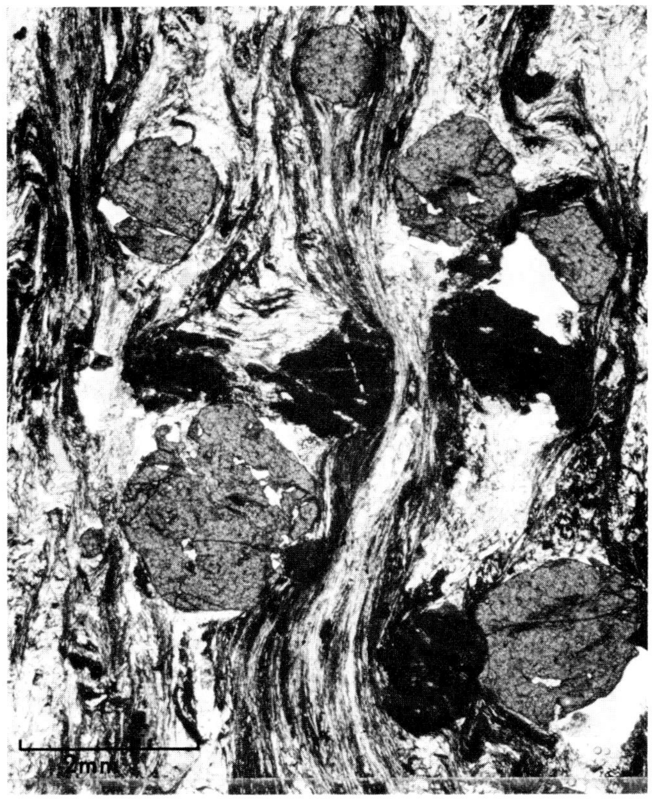

Phot. 32. Graphitic biotite-garnet schists. The garnets are rotated and the biotite/sericite bands show helicitic folding. W Cändebi.

F.4.3. The Bumthang-Djüle La metasediments

Above the Tongsa gneiss uplift and its eastern continuation which follows to the E of the Cändebi carbonate zone we recognize the matasedimentary basin of Bumthang with carbonate intercalations to the N and NW and which are overlain by the next higher northern migmatitic gneiss unit. The softer pelitic metasediments are to some extent responsible for the wide and open depression of the Bumthang region.

Dominant here are two-mica garnet schists, some with rotated garnets with S-like quartz inclusions (phot. 33). The schists contain kyanite in the coarser varieties. Some zones of biotite-psammite gneisses follow the center. They contain intercalations of garnet amphibolites together with bands of calc-silicates rich in diopside. Towards the NW are marble layers with some dolomitic bands not unlike the Cändebi carbonates. They increase northwards, become steep, dip to the NE and continue into the steep NW/SE striking Djüle La belt. This is well exposed on the Djüle La, the type locality for this zone, between the Bumthang region and the upper Tongsa valley.

This same belt continues eastwards over the Paibai La into the upper Thang Chu valley, a

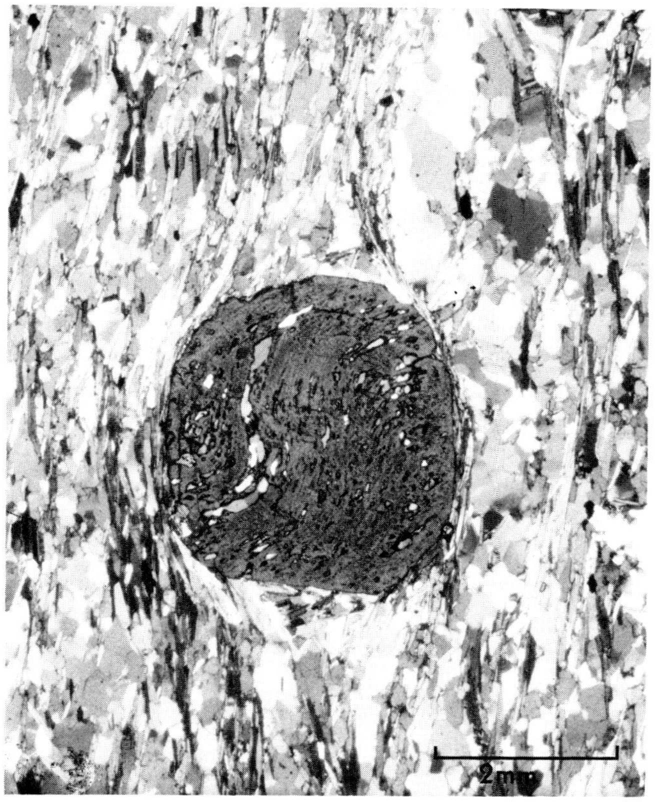

Phot. 33. Rotated garnets (S like quartz inclusions) in two mica garnet schist Bumthang region. The same rotated garnets occur in similar schists in the Paro region.

notheastern tributary of the Bumthang Chu and not to be confused with the Tang Chu from Central Bhutan. On the Paibai La outcrop sugary dolomitic marbles and thin layers of banded amphibolites. The marbles are overlain by conspicuous garnet bearing graphite-staurolite schists alternating with some biotite quartzites. The staurolites content increases above the famous Thowada rock monastry and corresponding rocks build up the mountain Pt. 4060 m to the N of the Gömpa. Here large staurolite crystals, rich in penetration twins, are embedded in graphitic biotite-(muscovite)-garnet schists with a marked green amphibolite layer (fig. 73).

Above and to the N follow flat, saucer like gneiss sheets of the northern crystalline thrust mass. The well bedded gneisses of the Takhtsang type have a very wide extension (fig. 74). In the southern Thang Chu valley, below the staurolite-marble zone, are exposed biotite-(muscovite) gneisses, amphibolites and kyanite-muscovite schists. This section overlies the gently northwards dipping coarse flaser type biotite augengneisses at the confluence ot Thang Chu and Bumthang Chu. They form the base of the Bumthang metasediments.

The actual Djüle La belt begins at Kakhtang near the Naspe Gömpa in the Chamka Chu valley (N of Bumthang) where the well bedded section is overthrust towards the S and SW by thick bodies of migmatitic biotite-alkalifeldspar gneisses belonging to the northern crystalline unit (fig. 75). The same thrust can be followed all along the NE side of the Djüle La belt though its actual trace is, as usual, not always easy to define. It is also responsible for the spectacular hot springs in the upper Tongsa valley.

At Kakhtang is a well exposed section of alternating coarse grained white marble bands, well bedded fine grained biotite-psammite gneisses and a spectacular zone of coarse grained garnet-staurolite-biotite-muscovite schists (phot. 34). Quartz veins rich in large kyanite crystals with muscovite and biotite occur locally. The staurolite crystals are several cm in size. Some of the crystals show poikiloblastic inclusions of quartz grains and some biotite (phot. 35, 36, 37, 38). The garnets are mostly free of inclusions and consists of an unaltered almandine. Sillimanite has not been observed in this zone, but is present in the Djüle La zone further to the NW as well as in the overlying gneiss body, which belongs to the Takhtsang type crystalline.

Towards the Djüle La the metasediments have steepened in dip (fig. 76, 77). In the Domdjen area, SE of Doli La, sillimanite appears in the schists. The staurolite has disappeared and we find garnet-sillimanite-biotite schists as well as gneisses. They are rich in sillimanite which occurs as larger crystal bundles of fibrolite. The idiomorphic garnets (almandine) are free of inclusions (phot. 77, 78, 79). We have here a good example of how the metamorphic isograde cuts an important structural trend. The carbonate rocks have increased in abundance and form several meters thick layers alternating with calc-silicate bands and psammite gneiss (fig. 78). They are medium grained and often white, yellowish and grey banded, the lighter coloured parts sugary and dolomitic. Sills and dykes of a pegmatitic garnet bearing muscovite-(biotite) granite locally cut the carbonate zone (fig. 79). They are related to a larger granite intrusion N of the Djüle La. The intrusive contact is vertical or steeply northwards dipping and in spite of its intrusive character is mostly concordant with the here equally steep dipping Djüle La zone (fig. 80). The biotite granite is medium to fine grained, mostly massive with inclusions of fine grained dark biotite-psammite rocks. Nebulitic to migmatitic schlieren occur locally. Contact metamorphism is minor and the high grade facies is a regional and not a local feature.

Over the pass, the Djüle La zone continues into the upper Tongsa valley (fig. 81). Here along its northern contact, where a secondary fault, related to the major thrust, is supposed to pass, we find spectacular hot spings. Just above the Tongsa or Mangde Chu occur a row of six springs, each with travertine deposits. The water has a temperature of 59°C, is sligthly sulfurous and contains CO_2 gas. Some springs issue already along the Tongsa Chu, still within marble bands with remarkable flow folding and intruded by aplitic garnet bearing tourmaline granite sills. Northwards, with a relative sharp contact begin thick bedded migmatitic biotite granite gneisses still with some muscovite (biotite) aplit

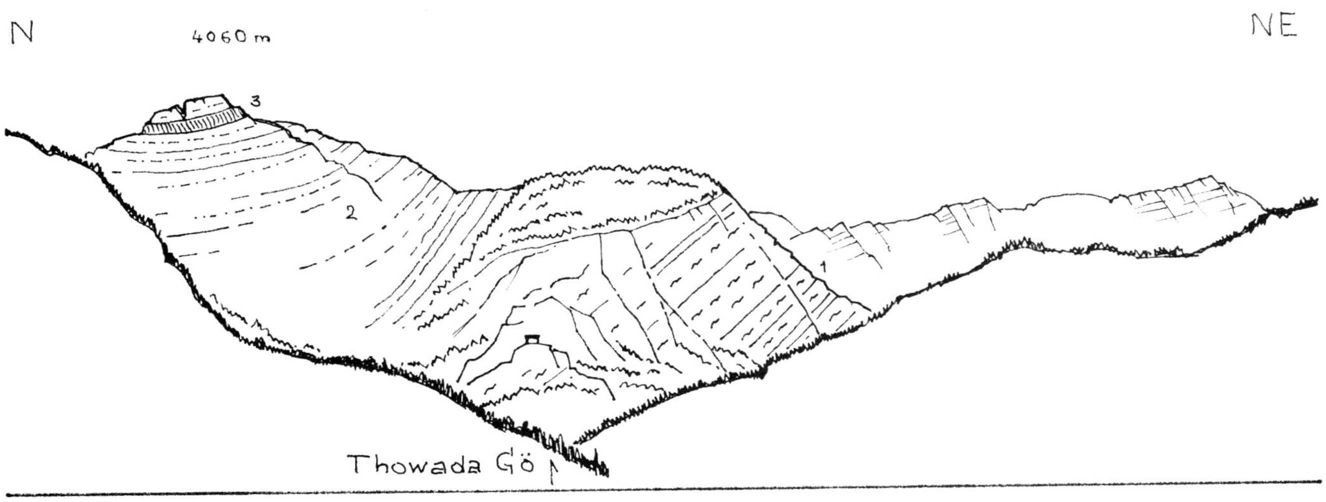

Fig. 73. The Thowada Gömpa region in Thang valley (NE Bumthang).
1 quartzitic biotite schists and gneisses, *2* garnet-biotite-staurolite schists, *3* amphibolite.

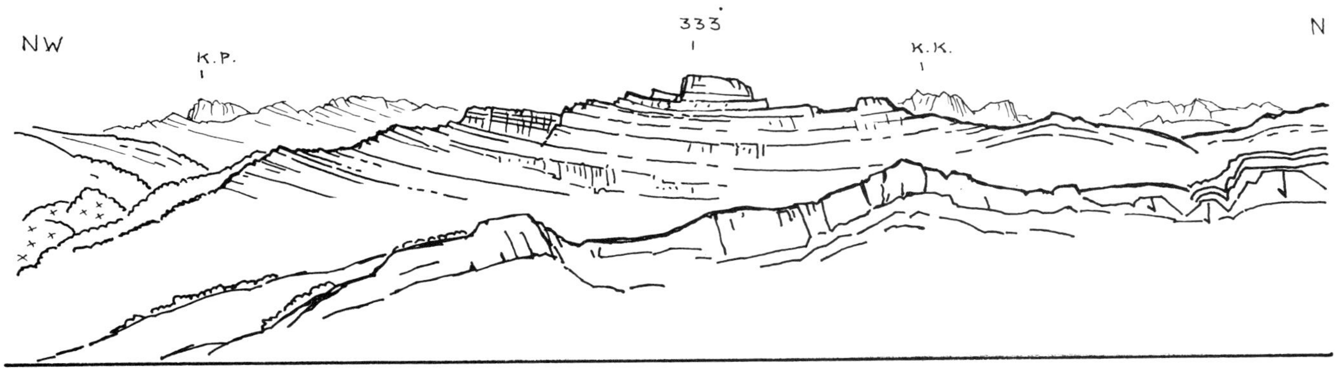

Fig. 74. General view to the N from Pt. 4060 m above Thowada Gömpa. Flat, saucer like, well bedded gneiss layers. *KP* = Kankar Pünzum, *KK* = Künla Kangri. To the right local tourmaline granite intrusions.

Fig. 75. The Chamka Chu gorge between Tsamba and Kakhtang.
1 Tsamba marbles with biotite-sillimanite gneisses on top, *2* biotite streiffengneiss, locally migmatized, *3* biotite granite (local granitization), *4* thin banded biotite gneisses, locally intensely folded, *5* migmatitic biotite gneisses with strong flow folding, *6* amphibolitic biotite gneisses, *7* psammite gneisses and biotite schists, *8* garnet-staurolite-biotite schists alternating with kyanite schists, calc-silicates and marbles (Kakhtang Zone).
A. Schematic cross section through Chamka Chu gorge showing old valley torso with fresh erosion along valley parallel fracture zones and faults.

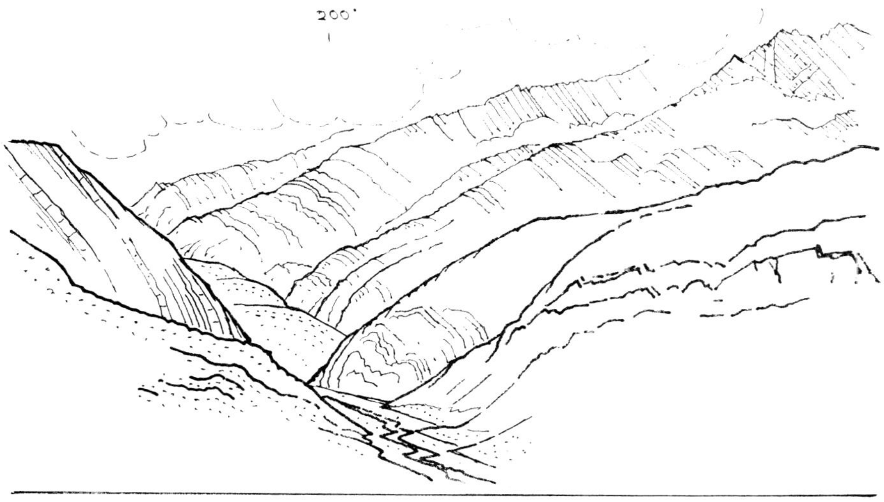

Fig. 76. The steep Djüle La zone seen from above Lamtang towards S. Note large moraines of Lingshi stage.

Fig. 77. The steep Djüle La zone in the upper Tongsa valley
1 garnet-biotite schists, *2* banded quartzites, *3* banded marbles and calc-schists, *4* aplite sills.

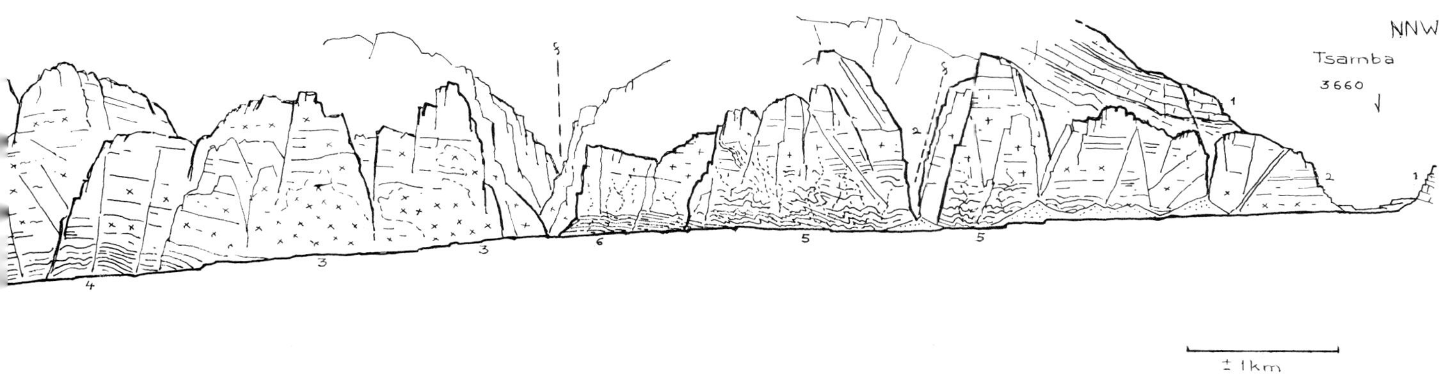

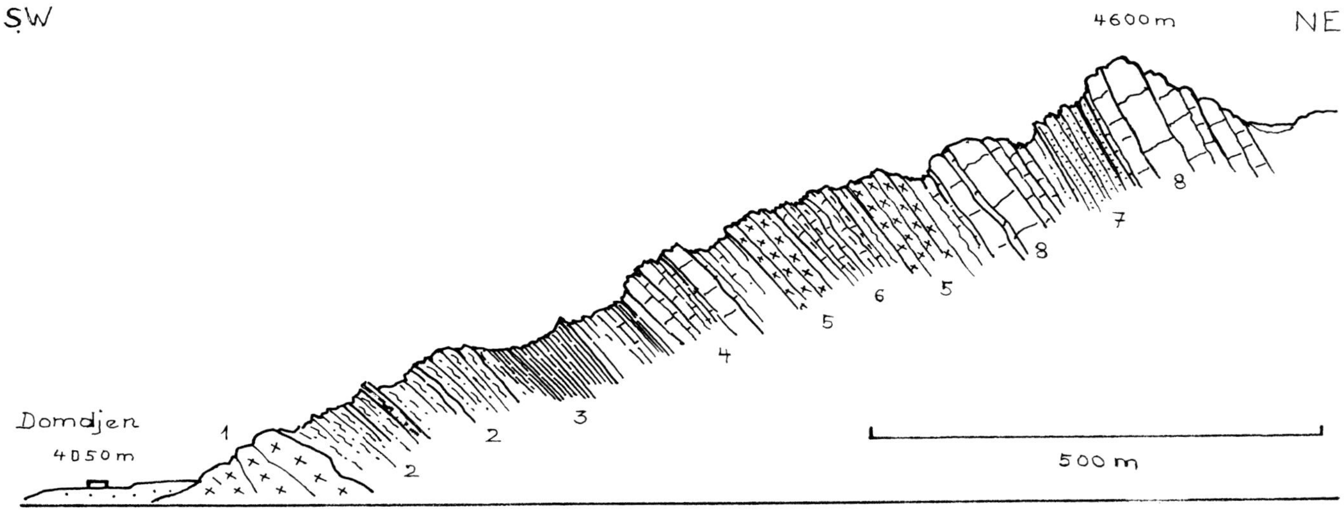

Fig. 78. The Djüle La zone at Domdjen.
1 garnet-muscovite-biotite gneiss, *2* garnet-biotite schists, *3* black graphitic slates, *4* white, medium grained well bedded marble, *5* well bedded biotite-psammite gneiss, *6* calc-schists, *7* well bedded white quartzites, *8* thick bedded coarse wollastonite marble with grammatite bands.

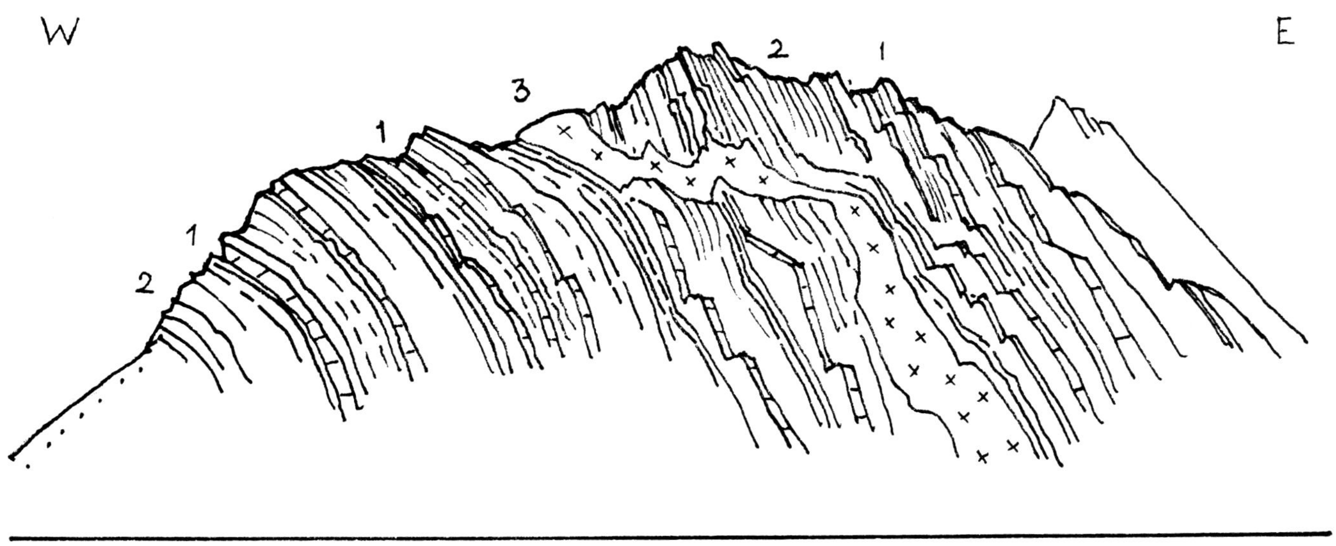

Fig. 79. Djüle La metasedimentary belt. SSE of Domdjen.
1 marbles and calc-silicates, *2* garnet schists to biotite-psammite gneisses, *3* crosscutting aplitic tourmaline granites.

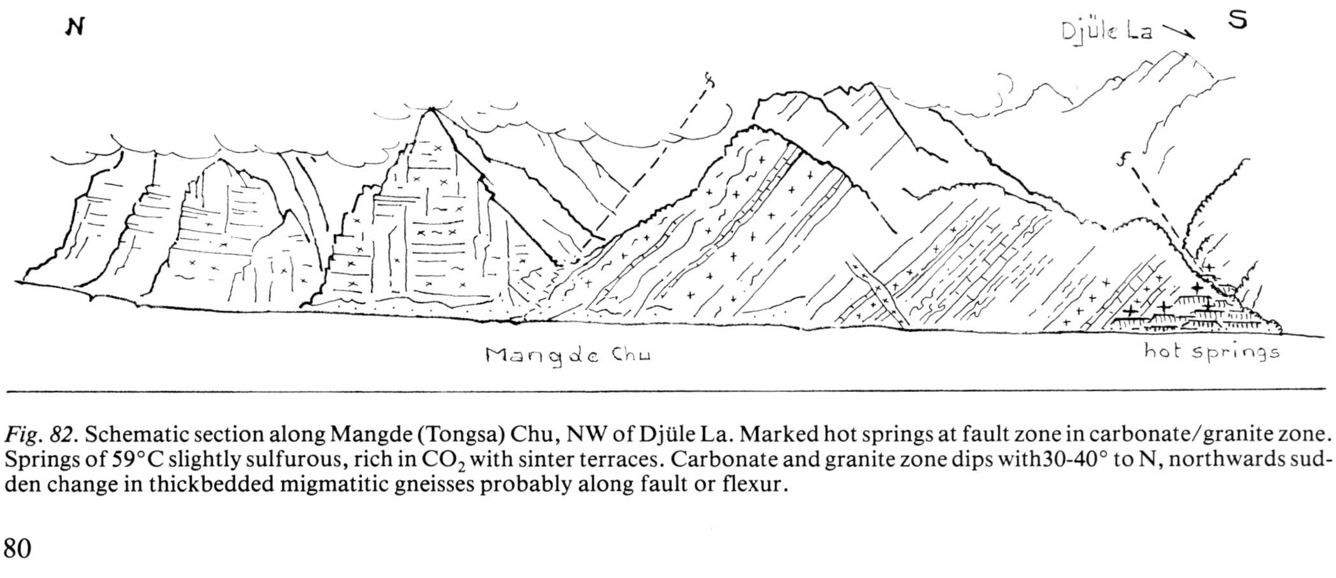

Fig. 82. Schematic section along Mangde (Tongsa) Chu, NW of Djüle La. Marked hot springs at fault zone in carbonate/granite zone. Springs of 59°C slightly sulfurous, rich in CO_2 with sinter terraces. Carbonate and granite zone dips with30-40° to N, northwards sudden change in thickbedded migmatitic gneisses probably along fault or flexur.

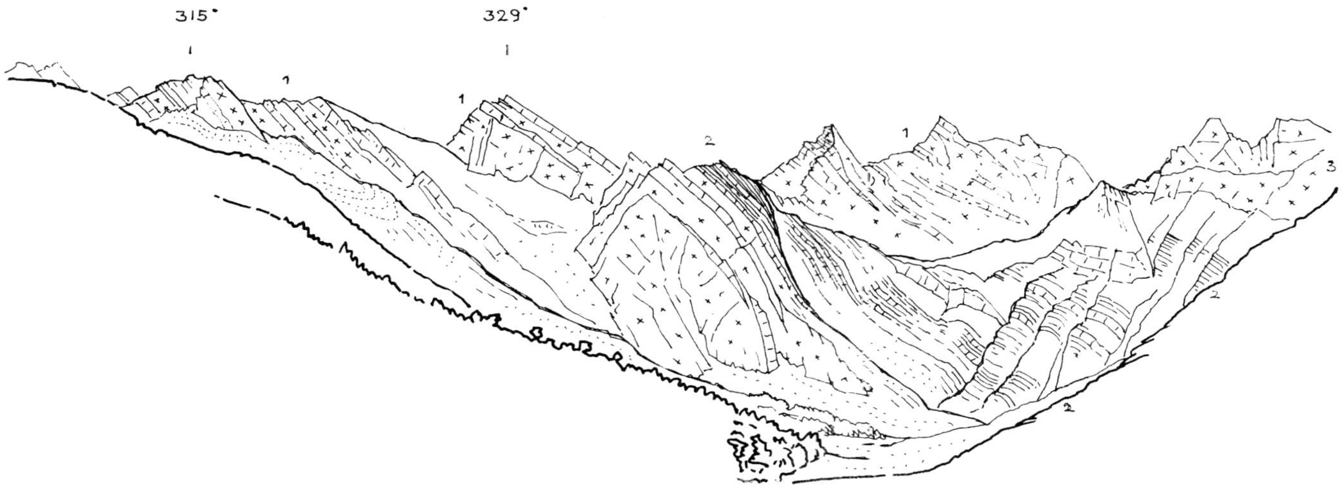

Fig. 80. The steep metamorphic sedimentary zone of Djüle La SE of Lamtang with intrusions of garnet bearing muscovite-granite. *1* marmorized limestones and dolomites in garnet bearing muscovite (biotite) granite, (often pegmatitic), *2* biotite garnet schists, black graphitic schists and calc-schists alternating with biotite-psammite gneiss, *3* homogeneous granite to granite gneisses.

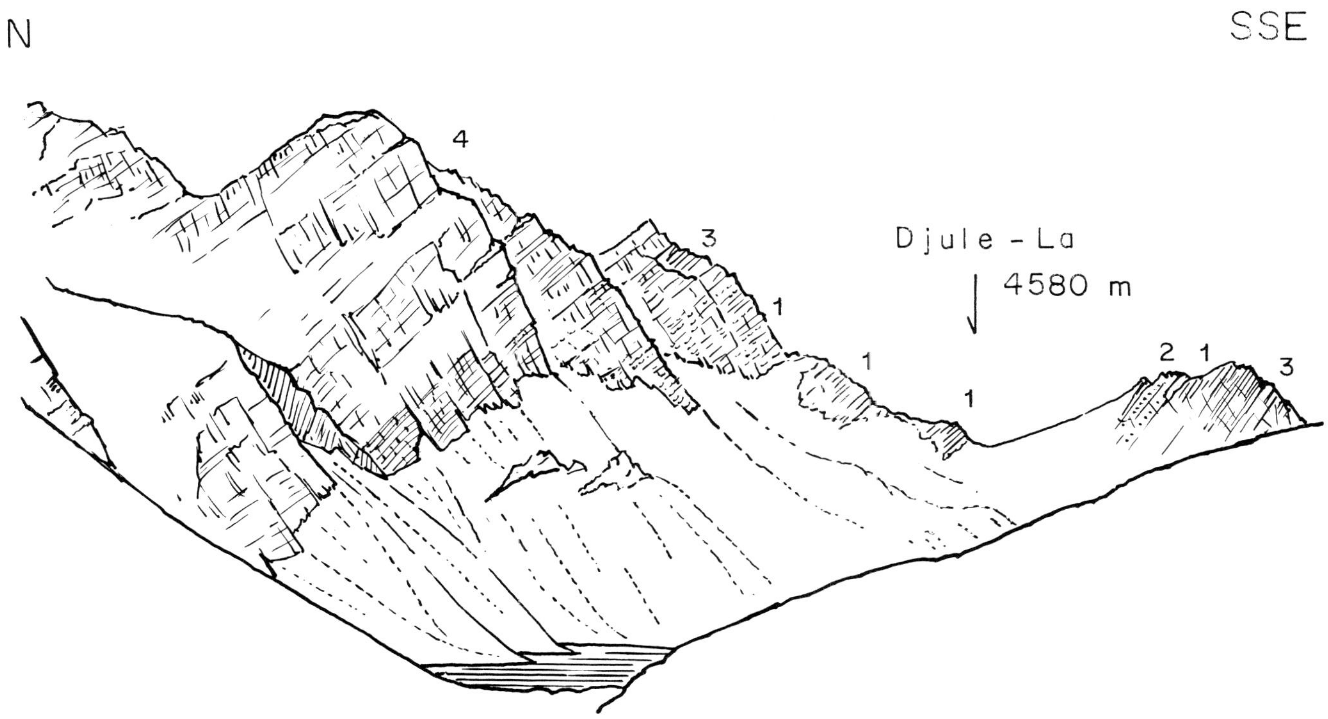

Fig. 81. The steeply north dipping Djüle La zone, west of the pass.
1 garnet-biotite schists, *2* banded quartzites, *3* banded marbles and calc-schists, *4* thick bedded dolomites and limestone marbles.

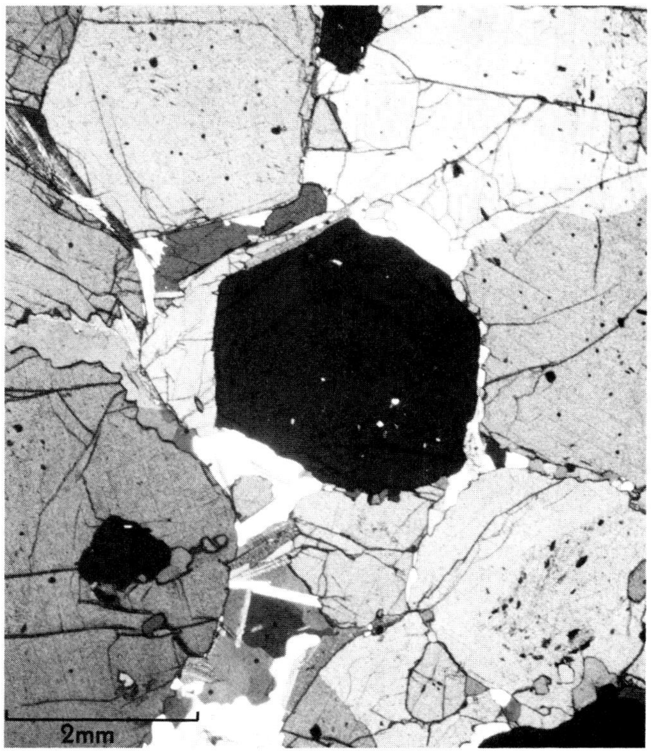

Phot. 34. Garnet-staurolite-biotite-muscovite schists. The large staurolite grains dominate with an idiomorphic garnet in the center. Kakhtang, northern Bumthang area.

Phot. 35. Poikiloblastic staurolite grains in biotite-muscovite, garnet bearing shists. Most of the inclusions are quartz with some biotite. Kakhtang, northern Bumthang.

Phot. 36. Staurolite with slightly S shaped aligned poikiloblasts and solid rim with crystal faces. From garnet-staurolite-biotite-muscovite schists, Kakhtang.

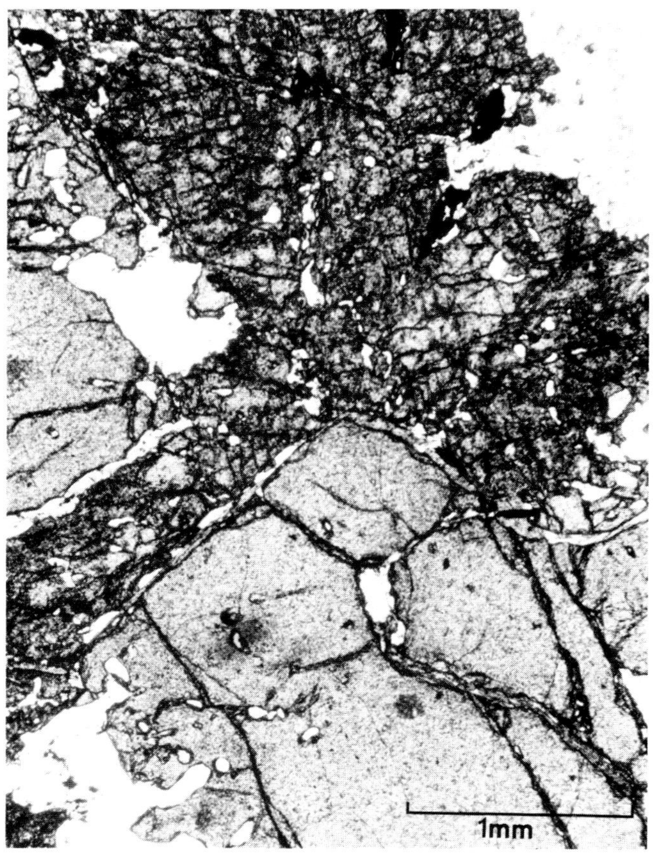

Phot. 37. Xenomorph poikiloblastic staurolite with garnets. Garnet-staurolit biotite-muscovite schists. Kakhtang.

Phot. 38. Poikiloblastic staurolite with solid rim (upper left) and garnet in contact with muscovite and biotite. Schists from Kakhtang.

Phot. 39. Phlogopite-olivine marble. The large olivines are partly serpentinized with a black iron ore rim. Kesha La, Lunana area.

granite intrusions. The latter resemble the younger granites however except for the absence of typical tourmaline. These gneisses grade into the northern gneiss mass which becomes suddenly much flatter and finally dips even gently to the S. The assumed thrusts seem to pass along the southern gneiss contact, but its outline is less sharp than further to the SE (fig. 82).

Continuing to the NW, the Djüle La zone is somewhat less steep and well bedded. After crossing the upper Tongsa valley, carbonate rocks reappear in the wild mountains of the Zaga La which leads into the eastern Lunana area. Here dolomite and limestone marbles form cliffs several 100 m high in this rugged mountain area. They are gently folded but severely faulted and dip in the average towards the NW which means below the gneisses of the Lunana region. Intercalated are banded biotite-psammite gneisses and at the edges of the marbles often calc-silicate bands rich in diopside with bytownite plagioclase. They occur together with boudinaged amphibolites. Marbles and gneisses are profusely intruded by fine to medium grained young tourmaline granites with muscovite and biotite. They show contact metamorphism with the carbonates yielding larger vesuvianites. This contact metamorphism overprints the high regional metamorphism since the intrusions of the granites cut all visible structures and are not tectonically affected.

F.4.4. The Lunana metasediments

From Zaga La westwards the marbles are interrupted by large tourmaline granite intrusions and by structural complication of the southern Lunana region. Corresponding zones reappear however in the central part of Lunana along the higher ridge to the S and SE of the valley. A lower horizon is well developed at the Kesha La (fig. 83) and from there runs to the N and NE along the eastern flank of the large Domchetang glacier (fig. 84). The westernmost belt was found just W of Karakachu La in the Rodophu glacier region just W of Lunana.

At the Kesha La a thick section of garnet-biotite-psammite gneiss with strikingly skeletal garnets is overlain by banded phlogopite-olivine marble. The olivines within the larger calcite crystals are partly serpentinized with a black iron ore rim (phot. 39). The phlogopites are very pale in colour and concentrated in layers. This striking rock is followed by diopside-calcite marbles and thicker bedded and coarser grained white calcite-phlogopite marbles with some small rounded diopside grains. They are overlain by banded biotite amphibolites rich in sphene and with basic plagioclase. It is interesting to note that some biotite zones are partly altered into chlorite. This retrograde metamorphosed horizon is rather unexpected. It may be related to some tec-

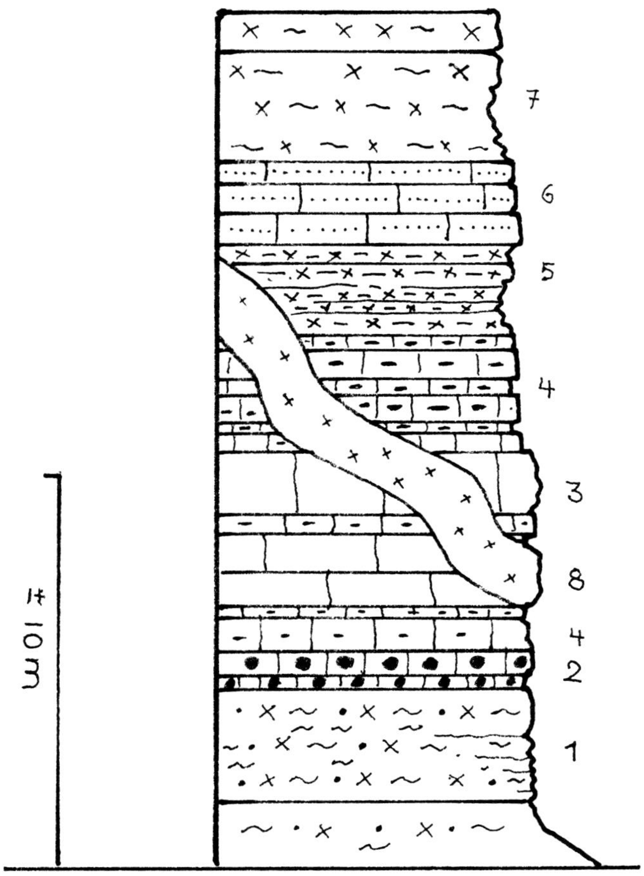

Fig. 83. Detailed section of metasediments at Kesha La, western Lunana.
1 garnet biotite gneiss,
2 olivine-phlogopite marble,
3 coarse white marble,
4 biotite-diopside calc-silicate,
5 biotite-psammite gneiss,
6 skapolite-diopside quartzite,
7 biotite-granite gneiss,
8 tourmaline aplite-granite dyke.

tonized contact with the overlying biotite-psammite gneiss. Any movement predates however the intrusion of the young aplite granite dykes, cutting through this section. The gneiss is again followed by a banded calc-silicate with diopside, scapolite, quartz and some sphene which shows no trace of retrograde effects. It is topped by a thick section of biotite psammite gneisses with intercalations of two mica granite gneisses. The Kesha La profile is only a part of the total carbonate zone which continues into the steep wall on the SE and E side of the Domchetang glacier where the layering is very well exposed (fig. 84). This zone may connect towards the W with the carbonate intercalations of the Karakachu La and Rodophu area passing N of the Gonju La, though the direct connection is not known. Again tourmaline granite intrusions which postdate the tectonics break the continuity.

Several carbonate bands can be recognized in the Karakachu La-Rodophu area (fig. 85). Just W of the pass within banded biotite gneisses we note a biotite amphibolite with labradorite and a brown-olive pleochroitic hornblende together with a white, medium grained diopside-calcite marble and layered calc-silicate bands. They consist of a zone of very fresh bytownite-plagioclase, the latter with most interesting melting reaction, not unlike some myrmekitic replacement. While the diopside-bytownite band shows no reaction at all, the melting reaction is restricted to the biotite-bytownite zone. K-feldspar seems to grow within the basic plagioclase. CO_2 and K respectively may be responsible for this striking differential reaction within the same slide. We certainly must assume high temperatures for this metamorphic zone.

On the Tsome La, leading into the Rodophu area, the carbonate bands are intercalated in strongly boudinaged migmatitic biotite gneisses with biotite-

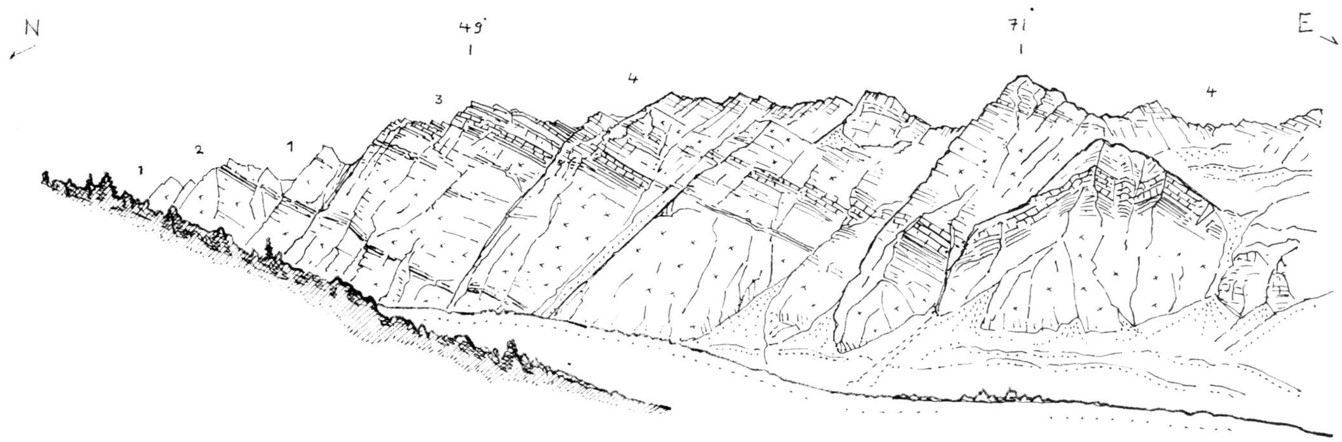

Fig. 84. The Domchetang valley, western Lunana. Northwestwards rising carbonate zone with biotite psammite gneiss in biotite migmatite gneisses.
1 migmatite gneisses with intrusions of tourmaline granites, *2* rusty weathered biotite schists (lower division), *3* banded marbles and calc-silicates in biotite psammite gneisses, *4* well bedded migmatitic biotite gneisses with some smaller calc-silicate bands. Locally intrusions of tourmaline granite.

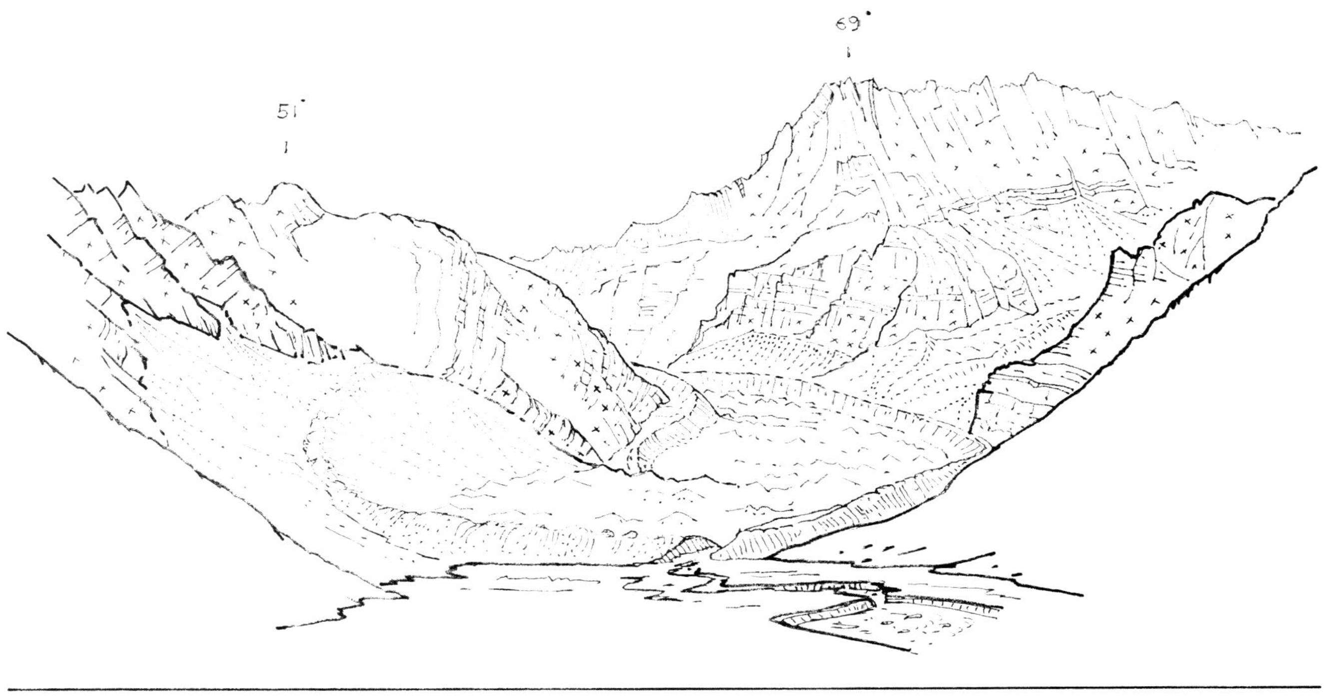

Fig. 85. The northeastern end of Rodophu valley, eastern side branch of upper Mo Chu. Zone of marbles and calc-silicate layers within banded garnet-biotite gneisses and striated and partly boudinaged migmatites. Locally intrusions of tourmaline granite, increasing eastwards and upwards. Note two well developed major glacial stages and young break through of dammed glacial lake.

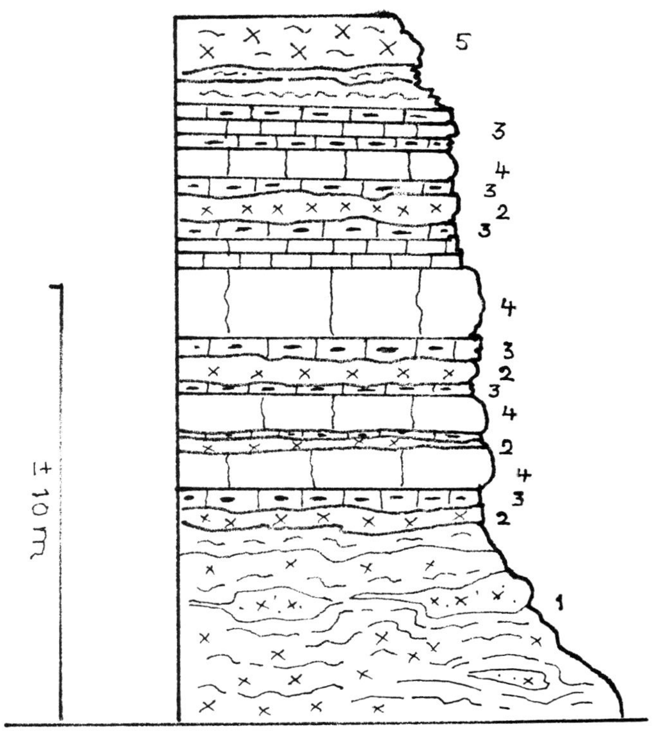

Fig. 86. Detailed section of metasediments on Tsome La (W of Lunana).
1 boudinaged migmatites in biotite psammite gneiss,
2 aplite granite sills,
3 banded calc-silicates,
4 coarse, spatic marble,
5 coarse, pegmatitic granite gneiss.

psammite gneiss layers (fig. 86). We recognize an alternation of thin calc-silicate bands and coarse white marbles intruded by sills of pegmatitic granite very poor in mica. The marbles contain diopside and scapolite. In the calc-silicate, free of calcite, the basic plagioclase changes into scapolite.

From the Rodophu area we see no continuation of the Djüle La type carbonate zone. To the W and SW we find the large migmatitic gneiss belts of the Masang Kang where carbonate zones are not known but where thin lenses of garnet amphibolite with ultramafic rocks are intercalated. Only further to the S, along the Mo Chu and S of Gaza Dzong do conspicuous carbonate zones within the gneisses outcrop again. They may extend into the Punakha region where they finally border the Tang Chu basin to the SW (see Geol. Map Plate I). Between Gaza Dzong and Punakha Dzong we recognize three main zones which consist mostly of calc-silicate bands, marbles and sills of tourmaline granite, a reduced section though not unlike the outcrops along the S wall of Chomolhari. At Tamji (S of Gaza Dzong) the marbles seem to have originated from a calcarenite, now rich in diopside but still with rounded quartz and feldspar grains within the calcite. The Tamji belt can be followed to the SE into the region of Tsephu, on the southwestern trail to Lunana. In the gorge below the village, marble and calc-silicate bands are intercalated between two mica augengneisses and biotite-psammite gneisses. The marbles as well as the calc-silicate contain diopside and some orthoclases together with scapolite. They are intruded by a sill of

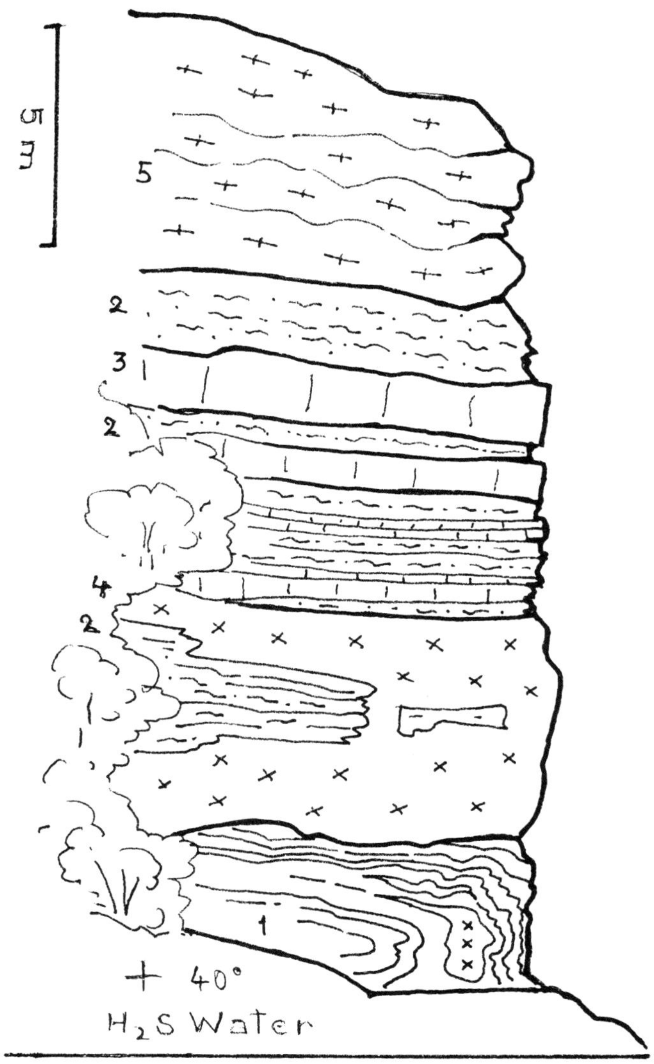

Fig. 87. Marble and calc-silicate section at hot springs, S below Tsephu (E of Mo Chu).
1 folded biotite psammite gneiss with lense of aplite granite,
2 banded biotite psammite gneisses to schists,
3 marble and calc-silicates alternating with psammite gneisses,
4 garnet aplite to pegmatite (late intrusion),
5 large augengneiss, same type as Tsephu with 10 cm K-feldspars (see fig. 111).

garnet bearing aplite granite and some garnet pegmatites. Gneiss xenoliths are visible in the granite (fig. 87). From the carbonate zone issue some hot springs with sulfur water and temperatures varying from 33° to 39° C. In the Punakha region the marbles are medium to coarse grained and always more or less rich in diopsides.

F.4.5. The northeastern metasediments

In the rugged and little known country of northeastern Bhutan, highly metamorphosed carbonate zones are common, but generally they coincide with extensive intrusions of the youngest tourmaline granites and apart from regional metamorphism the overprinted younger contact metamorphism plays here an additional role. I have studied them between the 7550 m high Kankarpünzum, the highest mountain of Bhutan and the legendary Sengge Dzong area in the E.

N of the Djüle La we already mentioned (fig. 80, 81) larger biotite granites intruding the metasediments with its carbonate belts. Further to the N, at the Doli La, the migmatitic granite which is different from and older than the tourmaline bearing leucogranites, intrudes N dipping biotite-psammite gneisses. They are overlain by a larger mass of lightgrey marbles which I called the Tsamba marbles. The latter increase in thickness and form wild summits intruded by an intricate system of pegmatitic dykes (fig. 90, 91; phot. 40). The base of the carbonate rocks is characterized by calc-silicate bands, while the higher parts turn into massive, mostly white, and often coarsely sugary marbles. The pegmatites contain some muscovite and garnet. We can follow the wild marble mountains to the ENE into the Chamka Chu valley where the marbles cross the valley at the beautiful place of Tsamba, the type locality for this northern calcareous zone. Here the marbles are about 500 m thick and consist of thick bedded layers with thinner bedded calc-schists. Flow features are frequent in the thicker beds. The marbles are topped by biotite-sillimanite gneisses with large sillimanites showing fibrolitic borders (fig. 92). Their regional dip is to the NNW.

Fig. 88. Djüle La steep zone in contact with Domdjen granite, the latter with pegmatitic dykes.

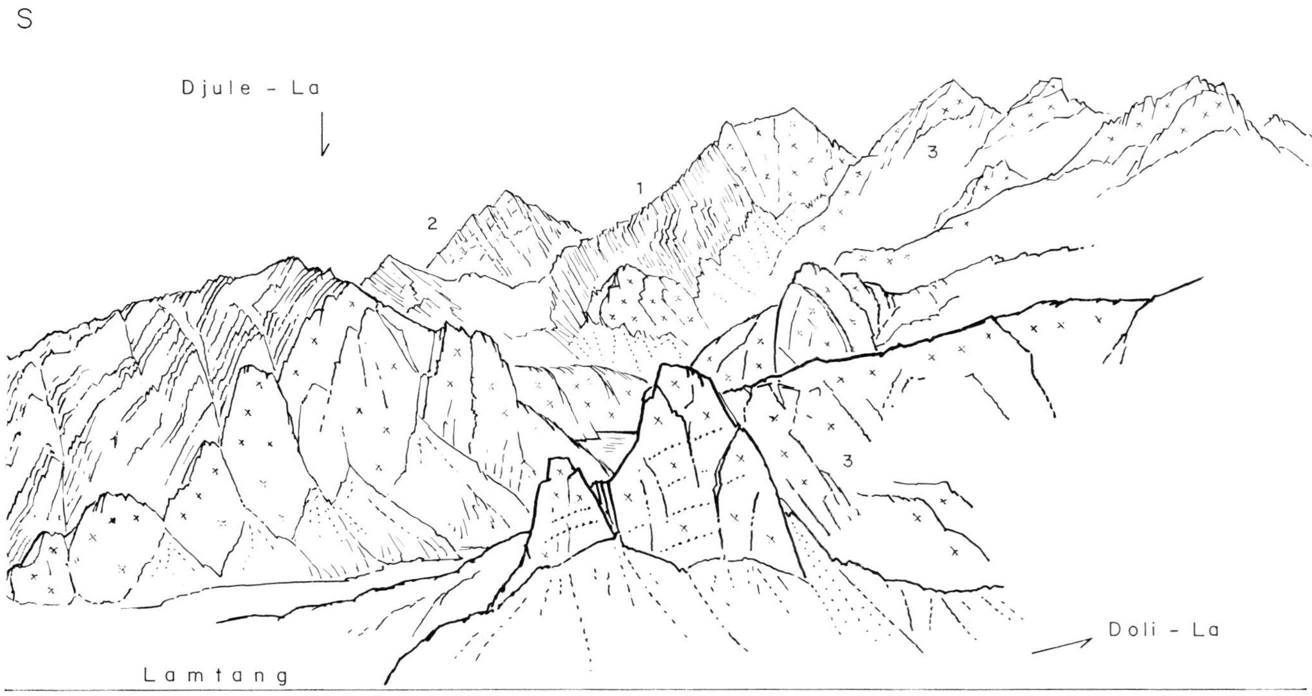

Fig. 89. The Domdjen biotite granite on the Djüle La zone (Lamtang valley).
1 biotite psammite schists, *2* marble zone, *3* biotite (muscovite) granite with pegmatite dykes (...).

Continuing upstream, after passing biotite-psammite gneisses with some migmatitic zones which cover the Tsamba marbles, we enter a most complex section where psammite gneisses, and higher up, marble and calc-silicate layers are intruded by sills, dykes and larger masses of tourmaline granite, already belonging to the large granite mass of the Monlakarchung La at the Tibetan border (fig. 93). Most of the carbonate layers are spatic, grey dolomitic marbles alternating with coarser grained dolomites rich in phlogopite. The intercalated garnet-biotite-sillimanite gneisses and schists can form irregular xenoliths within the larger tourmaline granites. Gneisses, schists and marbles show a marked lineation as well as larger folds in the lower part of the section. The axis plunges with 25°–30° to the NNW, similar to the regional dip of the metasediments.

The same intrusive contact zone with its array of irregular dykes was studied in the upper Melunghi Chu valley, which at Tsamba branches form the main valley in a northwesternly direction. In the headwaters of the Melunghi Chu along the Tibetan border we find the spectacular range of the Melunghi Kang

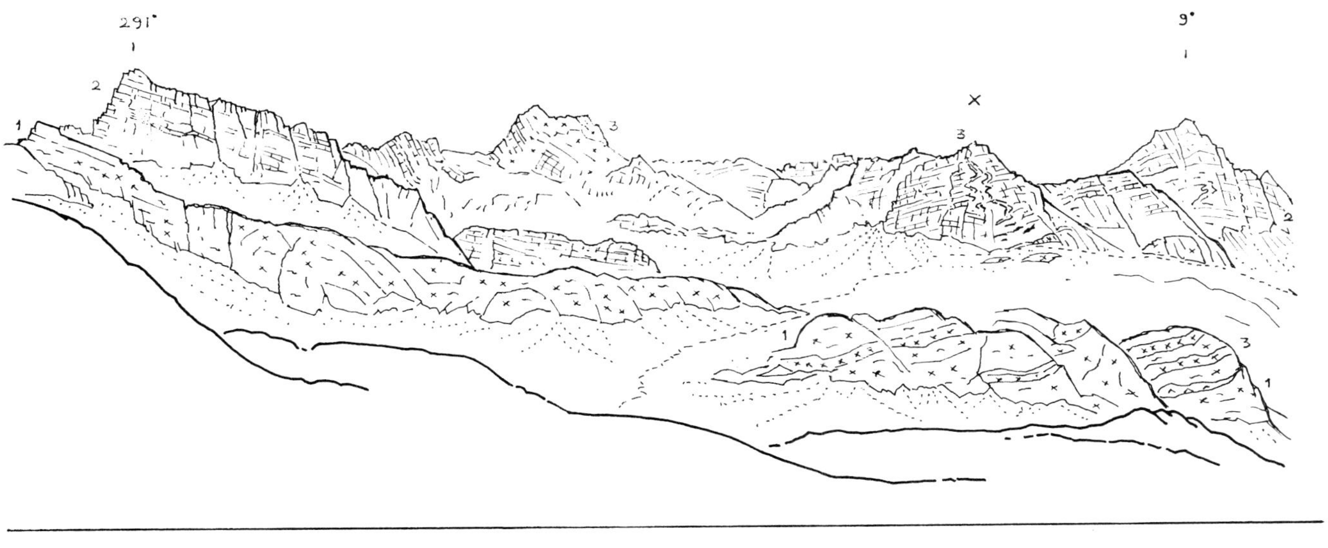

Fig. 90. View from Doli la towards N with Tsamba marble zone.
1 biotite granite gneiss to granite with biotite-psammite gneiss layers, *2* calc-silicate zones alternating with biotite-psammite gneiss to biotite-garnet schists grading upwards into thick bedded marbles (Tsamba type), *3* pegmatitic muscovite granite intruding into gneisses and marbles, *X* detailed sketch of granite intrusions (fig. 91).

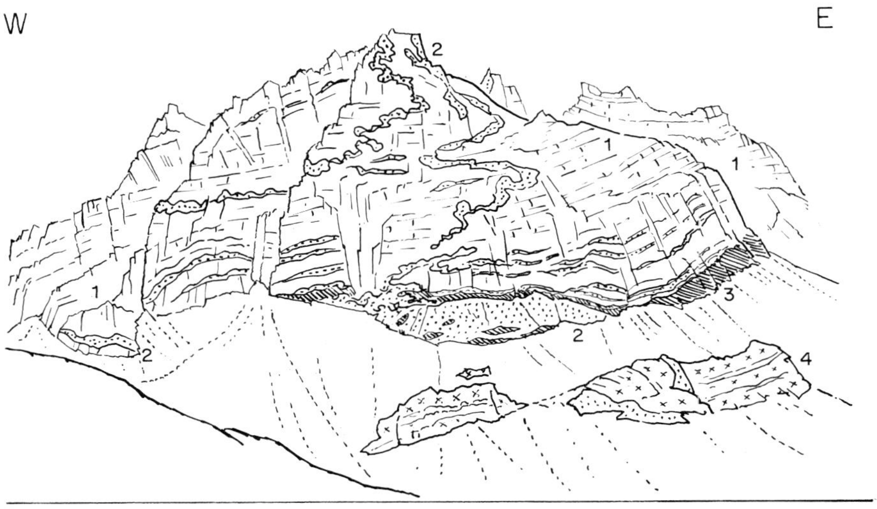

Fig. 91. The Tsamba marble zone with intricate system of pegmatitic dykes. North of Doli La, (WSW Tsamba).
1 gray and white marbles, *2* pegmatites (with some aplititic granite dykes and lenses), *3* biotite-psammite gneisses, *4* muscovite-biotite granite-gneiss.

(phot. 41) which continues directly into the Kankarpünzum peak (fig. 94). The summit of Melunghi Kang consists of bedded biotite-psammite gneisses with sill-like intrusions of tourmaline granites. The lower steep flanks and the ridge towards Kankarpünzum display masses of strikingly white tourmaline granite which continues into the Kankarpünzum peak. In some of the well exposed, nearly vertical granite cliffs are masses of xenoliths, mostly of «undigested» but sharply delineated biotite gneisses. The marbles form constant horizons and seem more resistant to the intrusion than the gneisses, a fact also known from the Nanga Parbat in the westernmost Himalaya (*Misch*, 1936). Below Kankarpünzum, on its SE side, the contact of the tourmaline granite with marble zones is well exposed, displaying a wide selection of contact rocks, comparable to the classical contact phenomena of the Tertiary Alpine Bergell and Adamello massifs. Of special interest are vesuvianite-garnet-diopside marbles rich in wollastonite and with vesuvianite crystals reaching a size of 2 by 10 cm. Thicker bedded white calcite marbles are often very rich in phlogopite. Compared to the southern section N of Tsamba, dolomite is practically absent. Along the upper Melunghi Chu valley, the marbles form a rather continuous zone of peaks following to the S of the valley (fig. 94). Further downstream they cross the valley and connect directly with the Tsamba marbles.

The tourmaline granite intrusion increases to the N, and towards the Monlakarchung La at the Tibetan border all the head valleys and the border range consist of young massive tourmaline granite (see later).

To the E of the Tsamba-Melunghi Chu region the tourmaline granites extend further to the S and cut the eastern continuation of the Tsamba marble belt just W of the Zakar La region (fig. 95). In this wild area, between the Melunghi Chu and the large Kuru Chu valley in the E the tourmaline granites extend up to the Tibetan border. Only further to the E, before reaching the actual Kuru Chu valley slope do we cross the easten intrusive contact of the granites and enter psammite gneiss sections with about four major carbonate zones (see Geol. Map. Plate I).

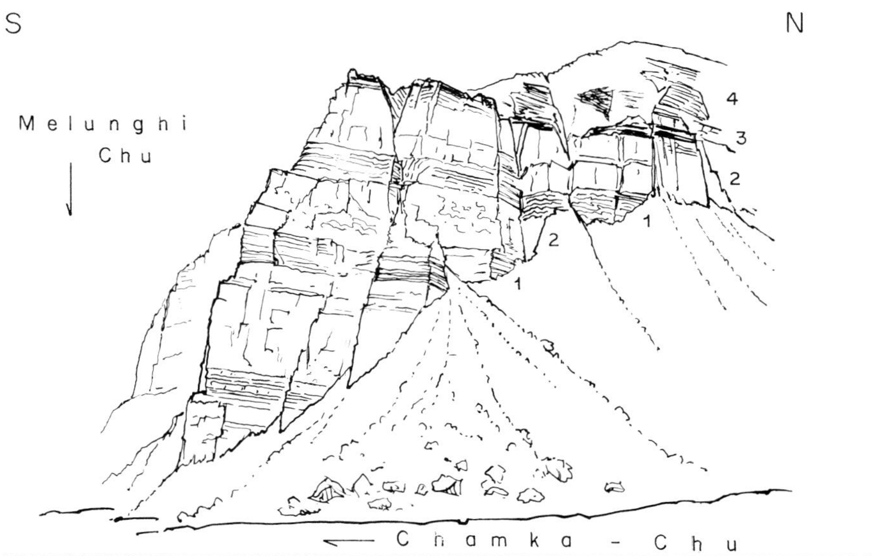

Fig. 92. The Tsamba marble type locality at Tsamba (Chamka Chu valley).
1 calc schists, *2* thick bedded marbles, often with flowing structures, *3* alternating marble bands and biotite schists, *4* biotite- sillimanite schists to gneisses.

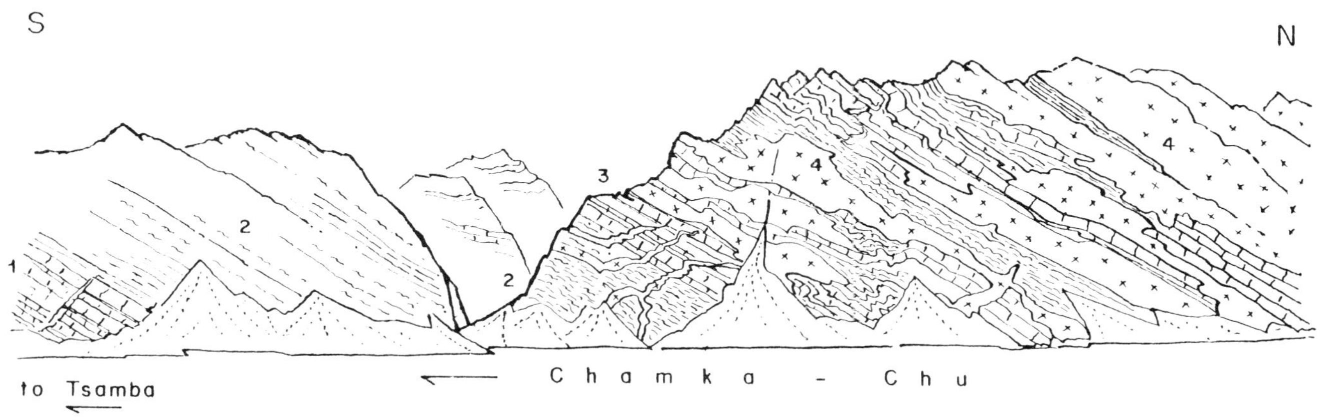

Fig. 93. The southern contact zone of the Monlakarchung leucogranite, (Chamka Chu valley).
1 Tsamba marbles, *2* biotite-sillimanite psammitegneiss, *3* dolomitic marbles alternating with *4* sills, lenses and dykes of tourmaline bearing leucogranite, the latter increasing northwards into the main Monlakarchung granite body.

Phot. 40. Tsamba marbles intruded by dykes and sills of pegmatitic granites. N of Doli La (see also fig. 91).

Fig. 94. The granite and marble mountains of the upper Melunghi Chu. Large development of the Thanza stage moraines. *1* leucogranites, *2* biotite psammite gneiss, *3* limestone and dolomite marbles of the Tsamba type.

The contact is well exposed in the Dermalung valley (fig. 96). The tourmaline granite cuts with a very irregular discordant contact well bedded NNW dipping psammite gneisses which also occur as xenoliths near the contact zone. The xenoliths are rotated within the granite. Granite apophyses enter the gneisses for several 100 m. The latter consist of biotite gneisses with some olive brown pleochroitic biotites and andesine. They grade into migmatitic biotite gneisses where the biotites are concentrated along the neosome. Intercalated are some fine bedded amphibolites with a middle basic plagioclase. Garnets and staurolite, common in other similar sections elsewhere are missing. Further to the SSE we meet the first carbonate bands. Striking is a very coarse white and pure calcite marble with crystals over 1 cm long. They emit a peculiar fetid smell when hammered. Below the marble follow alternations of phlogopite-diopside marbles, diopside marbles with calcite and basic plagioclase as well as some scapolite and an actual diopside fels with only very few calcite grains. It is bordered by fine grained diopside-hornblende gneisses rich in sphene. A well bedded quartzite divides the carbonate into an upper and lower zone. The lower zone continues over the Sergong La and reaches the Kuru Chu valley already beyond the impassable gorge (fig. 97). Towards the pass the carbonates are cut by pegmatitic dykes and

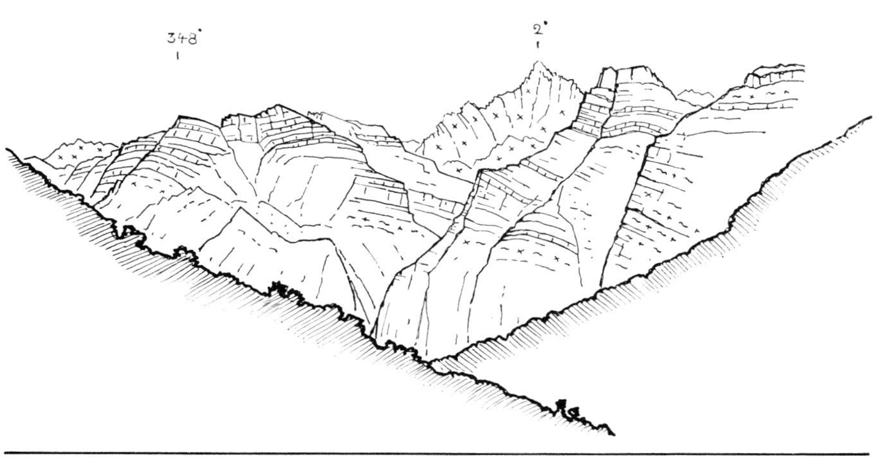

Fig. 95. Eastwards continuation of the Tsamba marble zone in Gomtang valley WSW of Zakhar La. Base migmatitic gneisses, further northwards intruded by tourmaline granites.

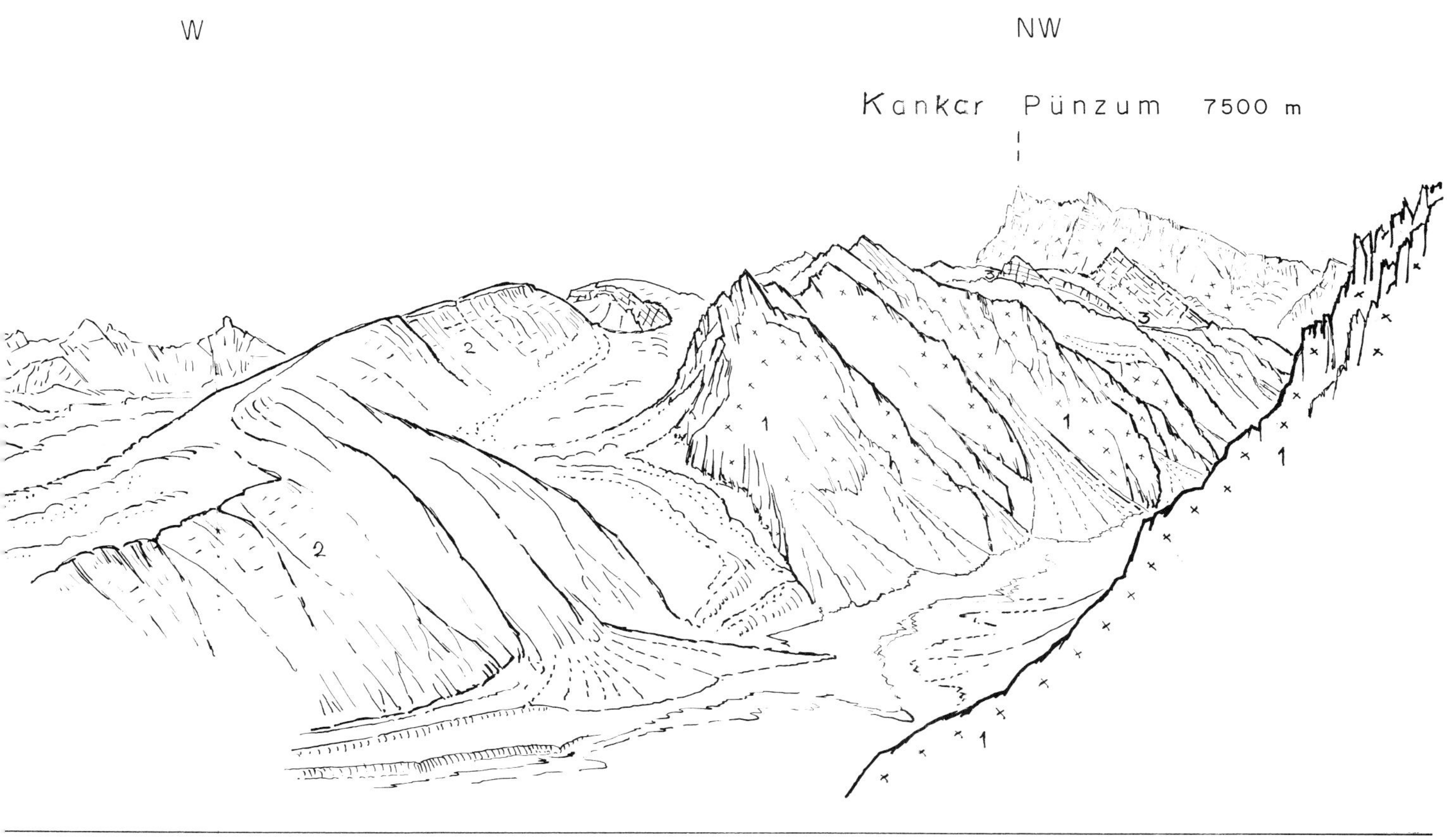

tourmaline granite sills, alternating with well bedded calc-silicates, psammite gneisses and bands of hornblende-diopside gneisses rich in sphene. Contrasting with this lower well bedded carbonate section, the upper carbonate layers are strongly folded with fold axis dipping with 20° to the NNE, coinciding with the dip of the bedding. It is interesting to note that the folding is mostly restricted to the carbonate zone. Flow folding seems to be the rule. No dolomites are present in these carbonate horizons in contrast with the sections N of Tsamba.

Further down the Kuru Chu at the village of Thunkar a second carbonate belt with white marbles and calc-silicate layers crosses the valley. They occur in garnet bearing biotite-muscovite gneisses which further downstream contain sillimanite. At the same time some of the biotites are chloritized, suggesting an incipient retrograde metamorphic overprint. This

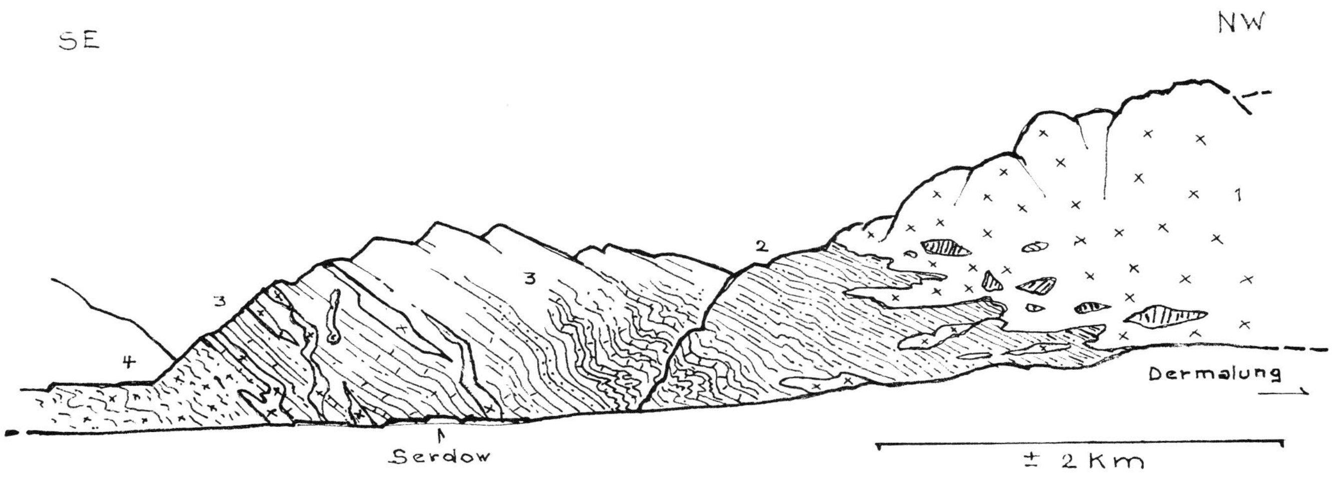

Fig. 96. Intrusive contact of tourmaline granite (*1*) with well bedded biotite gneisses and schists (*2*) and calc-silicate bands with marble layers (*3*). The latter and thin quartzites show intense flow-folding. Below follow migmatized biotite gneisses (*4*). Along western side-river of Kuru Chu between Ora La and Takpang.

Phot. 41. Melunghi Kang, SE wall. Summit shows banded biotite gneisses with leucogranite sills while the main wall consists of leucogranite rich in biotite gneiss xenoliths.

increases towards the Lhüntse Dzong region where we reach the N end of the Kuru Chu spur.

The same carbonate belt strikes westwards into the steep gorge of the Yalagang Chu just N of Neh. Here the carbonate rocks form a vertical cliff with a high waterfall and hot springs at the base (fig. 98). White, coarse, more or less phlogopitic marbles are followed by banded phlogopite-diopside-scapolite marbles with rounded quartz and plagioclase grains. They alternate with banded biotite-pyroxene gneisses rich in microcline which shows myrmekitic reactions with the plagioclase. Some of the pyroxenes are altered into a dark green hornblende. The carbonate belt is intercalated within sillimanite-biotite-garnet gneisses which constitute the main rock type. The frequent sillimanite forms larger bundles with fibrolitic edges. The almandine type garnets often contain inclusions of biotite, quartz, plagioclase and magnetitic ore. The whole section is traversed by pegmatitic muscovite-tourmaline-granite dykes and sills. The hot springs issuing from the basal coarse marbles are at a temperature of about 45°C and have a very slight sulfur taste. They seem otherwise free of carbonate and CO_2, since in spite of issuing from the carbonate level they do not form travertine deposits. They must originate from a deeper horizon than the carbonate rocks. In the head waters of the same valley, N of Zakar La, I found another hot spring at Pasalum. This spring issues from a fractured part of the here widespread young leucogranite. The 42°C hot water has a slight sulfur smell and has no travertine deposits. It seems very similar to the above mentioned springs N of Neh.

A third smaller carbonate belt runs through the

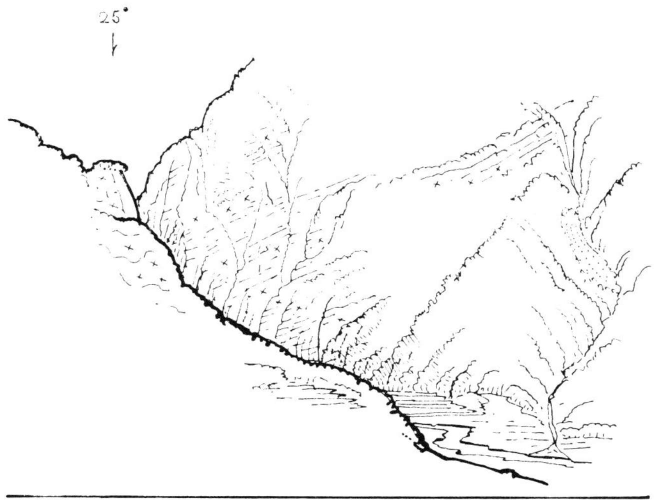

Fig. 97. The impassable Kuru Chu gorge N of Takpang, S of Tibetan border. N to NNW dipping migmatitic and banded biotite gneisses.

village of Neh and consists mostly of a coarse, whitish marble within psammite gneisses, cut by some pegmatitic veins. The psammite gneisses are underlain by very well bedded quartz-rich biotite gneisses and muscovite quartzites. They are strongly folded with fold axes dipping at 30° to the NE. They are of metasedimentary character in this section.

The fourth zone outcrops in the lower Koma Chu coming from the famous Sengge Dzong, above the confluence with the Kuru Chu and to the NE of Lhüntse Dzong (fig. 99). Above well bedded biotite gneisses with intercalated kyanite-biotite-muscovite-garnet schists follow several bands of calc-silicates and white marbles. The kyanites are 1–2 cm long

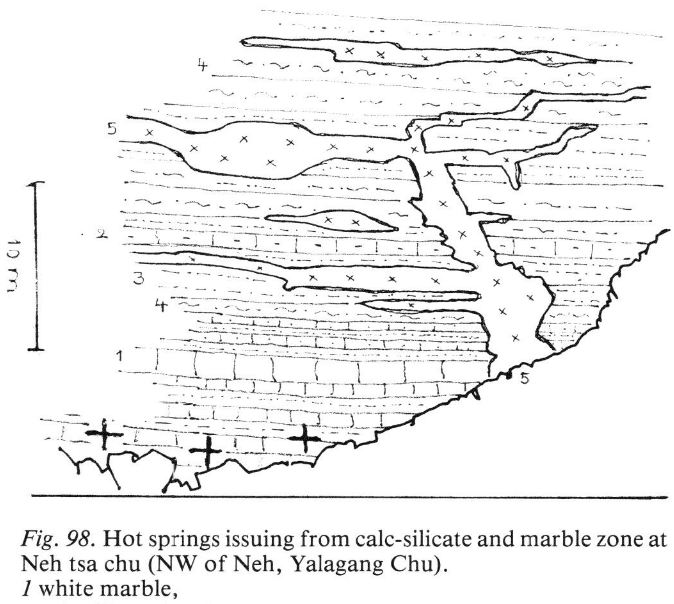

Fig. 98. Hot springs issuing from calc-silicate and marble zone at Neh tsa chu (NW of Neh, Yalagang Chu).
1 white marble,
2 calc-silicate bands,
3 biotite-psammite gneiss (banded),
4 biotite K-feldspar gneisses (striated),
5 pegmatitic muscovite-tourmaline-granite dykes and sills. From lower marble zone issue various hot springs, slightly sulfurous, but no carbonate in solution (no sinter) (40-45 °C).

idiomorphic crystals. The garnets are poikiloblastic with quartz and biotite as inclusions. Sillimanite is absent and the kyanite is a good indication for the metamorphic grade in this section. Some of the marbles expose sharp isoclinal folds within otherwise parallel bands. The folds dip in the same direction as the well bedded sediments with 30°–40° to the NW. Tourmaline pegmatites cut the marble zones.

Following the Koma Chu to the N we eventually cross the third and second carbonate belts already described from the Kuru Chu. Some of the marbles are white and grey banded and occur within more or less migmatitic biotite gneisses. They are overlain by conspicuous, thick bedded migmatitic biotite-granite gneisses, with the biotite concentrated along the neosome. Here, half way to Sengge Dzong outcrop garnet-sillimanite-biotite-granite gneisses, similar to the ones at Takhtsang type locality. They indicate and increase in the metamorphic grade which below does not surpass the kyanite facies. The granitic content of the rocks increases towards the Sengge Dzong region where migmatitic biotite granites rich in irregular biotite gneiss xenoliths form steep, massive peaks which surround this holy place with the many meditation caves and the famous Dzong. They contrast with a flat ground of lacustral sediments, dammed by rocky barrier which suggests an ancient glacial basin (fig. 100). The legend of Sengge Dzong speaks of an ancient lake from which a large «Phurbu» (ceremonial dagger) was retrieved. Some of the eroded schistose xenoliths within the granite have been taken as body and footprints of Yeshe Tshogyal, the mystic consort of Guru Rimpoche, who was the first worshipper of this place (*Olschak*, 1979). Here I collected a particularly interesting xenolith from the main granite just at one of the still active worshipping places (a cave called Dorji Dzong, where during my visit in December 1969 the Head Lama of Sengge Dzong was walled in for 2 years). It presents a remnant from the calc-silicate suite and consists predominantly of almandine garnet with biotite, andesine type plagioclase and some quartz. The garnet which makes up 70% of the rocks has a rim of quartz and plagioclase inclusions while the otherwise pure core contains some magnetite ore (phot. 76).

A last metasedimentary zone with thin marble bands and calc-silicates was observed at the 5000 m high Gwen La to the N of Sengge Dzong. The well bedded section exposes biotite gneiss-quartzites alternating with coarser biotite gneisses at the base and followed by calc-silicates with well banded pyroxene-hornblende gneisses in which the hornblende has developed from the pyroxene. They border coarse, grey marble layers which are topped by conspicuous actinolite-biotite-(muscovite) schists. The whole section is intruded by tourmaline bearing aplite granite sills with locally cross-cutting pegmatites.

Further to the N, the Gwen La metasediments are ovelain by steeply fractured granite gneisses which form the wild border range and strike towards the

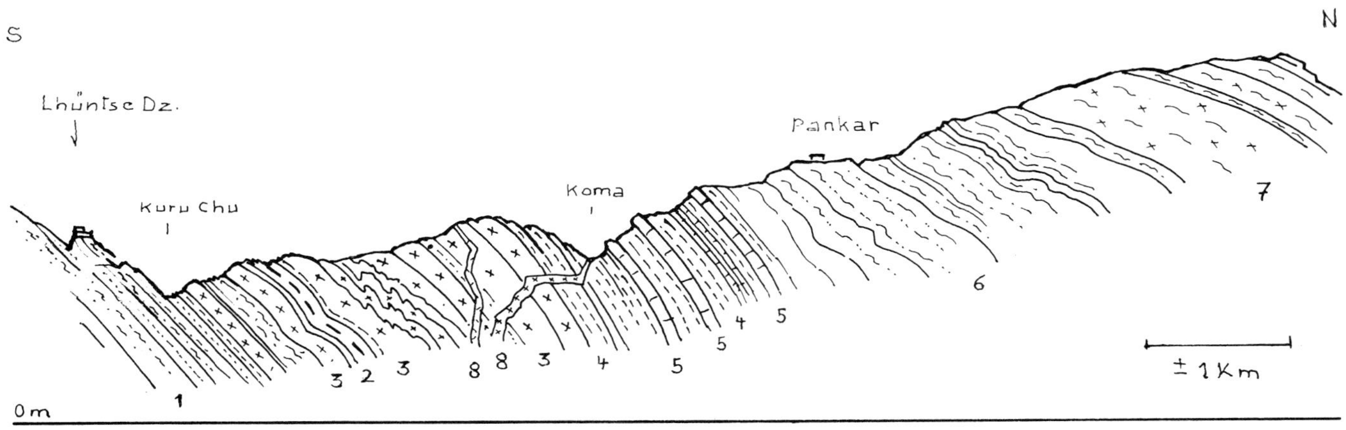

Fig. 99. The carbonate zone between Lhüntse Dzong and Pankar along the lower Koma Chu.
1 biotite gneiss and well bedded muscovite quartzites, *2* biotite-muscovite kyanite schists, *3* granite gneiss, *4* psammite gneiss, *5* well bedded marbles and calc-silicates, *6* garnet-muscovite-biotite gneiss, *7* migmatitic biotite gneiss, *8* crosscutting tourmaline pegmatites.

Bodo La (on the Tibetan border). Here the gneisses are intruded by a larger body of massive young leucogranites which further to the E intrude black slates of the northern ridges and summits of the little known Garula Kang group (fig. 51, 101; phot. 42). Looking from the Gwen La to the WNW one notes the wide depression of the upper Kuru Chu, already in Tibet, with rolling hills consisting of black slates. They are dominated further westwards by the granite and gneiss range of the 7600 m high Künla Kangri and its imposing NW continuation of high and still unknown glacier peaks (phot. 43).

Fig. 100. The valley of Sengge Dzong. Migmatite gneisses (*1*) with irregular granitized zones (*2*) and inclusions of calc-silicate lenses (similar to Takhtsang gneisses). View towards W.

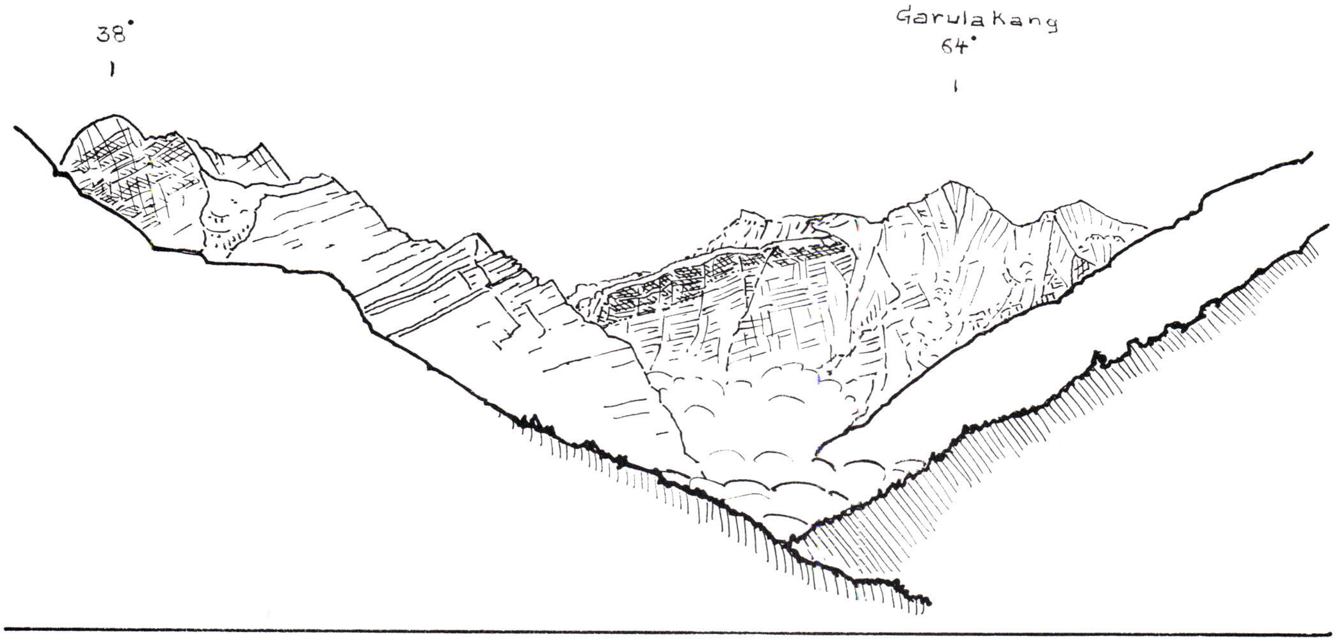

Fig. 101. Head valley NE Sengge Dzong with Garula Kang. Migmatitic gneisses grading upwards into well bedded biotite schists forming dark bands (bands of sedimentary series?) regional NW dip.

Phot. 42. Garula Kang S face, in the northeasternmost corner of the Bhutan Himalaya. Well bedded biotite schists and psammite gneisses, grading over biotite slates into the base of the Tethyan sediments of S Tibet.

Phot. 43. Künla Kangri with its 3 summits (7600 m) (left) and the newly discovered northeastern continuation of high, crystalline peaks. The central group shows northwards rising layers. The south ridge of the northern peak seems to expose a local synform. The whole range may form a pronounced northdirected crystalline spur. This range contrasts with the low rolling hills, mostly of black slates, outcropping along the Tibetan part of the Kuru Chu. In the foreground the gneiss range north of Gwen La. (see also Fig. 51) Note recent fracture line (f), probably caused by earthquake, (?). View towards W.

Fig. 102. The flat, somewhat dome like shaped (regional NW dip in the W and NE dip in the E) Takhtsang gneiss sheet N of Paro overlying the Paro metamorphic para series. The approximate contact, probably a recrystallised thrustzone is indicated by thrust symbols. Above kinzigitic Takhtsang gneisses with increasing meta-quartzites towards the top. Below Paro metamorphics, showing a gentler morphology.

F.5. The northern crystalline units

I have already alluded to the northern crystalline unit while discussing the metasediments which occur mostly in the form of carbonate intercalations within the wider gneiss bodies and are cut by the tourmaline granite intrusions. In the following I will discuss the main rock types of a more intrusive metamorphic origin, adding a special chapter on the young tourmaline leucogranites. I have also mentioned why I have chosen the Takhtsang rock monastery as a type locality for the northern crystalline rocks instead of the Thimphu region.

North of Paro in western Bhutan, the Paro metamorphics are overthrust with a more or less transitional contact by a large and extensive gneiss mass (fig. 102). Clinging to a near-vertical cliff of these gneisses is the famous Takhtsang monastery (phot. 44) which has been taken as the type locality for the Takhtsang gneisses. The contact between the Paro marbles and quartzites with the Takhtsang crystalline is shown in fig. 71. Along this section one notes a change from the mesometamorphic Paro marbles and quartzites to the sillimanite bearing gneisses. I place the major tectonic contact along conspicuous garnet amphibolites showing boudinage and marked zones of movement (phot. 45).

The Takhtsang rocks at the type locality consist of banded sillimanite-bearing garnet-biotite gneisses with white quartz and aplite seams (phot. 46). Locally the banding can become diffuse and the more massive gneisses appear to have been mobilized. There are all gradations to migmatitic types, with most of the original mineral content preserved. Within the gneisses occur eye-shaped lenses of fine grained calc-silicates, rich in garnet, diopside-augite, labradorite-bytownite feldspar but devoid of micas. Most of the garnets are net like relict grains. These peculiar concretions were interpreted as foot and hand prints of the founder of this place, the mystic consort of Padmasambhava or Guru Rimpoché, the «Fairy of Wisdom Yeshé Tshogyal, the reputed woman founder of holy places» (*Olschak*, 1979), who left her imprints in the rocks of Sengge Dzong. In both places we note how the historical-religious interpretation of preculiar rock formations lead to the foundation of famous monasteries. The whole aspect of the Takhtsang gneisses with its inclusions is strikingly similar to the Darjeeling gneisses; actually the similarity is even better between the Takhtsang gneisses and Darjeeling than with the Chasilaka gneisses further to the S, which seem to form a direct continuation with the Darjeeling gneiss of Sikkim.

Northwards, following the Paro Chu, well bedded quartzite horizons, mostly coarse grained and with translucent quartz crystals, occur intercalated in the northwards dipping Takhtsang gneisses (fig.

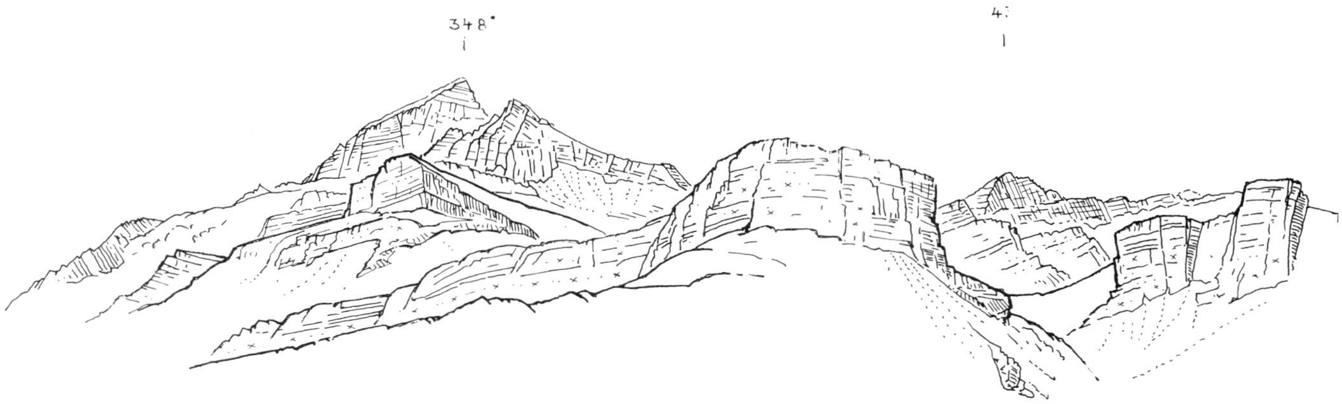

Fig. 103. The southern extension of the Chatorake range N of Paro. Regionally NNW dipping quartzitic well bedded gneisses. Seen freom N of Bhumda Gömpa.

Phot. 44. Takhtsang Monastery built on garnet-sillimanite-biotite gneisses. They are very thick bedded and dip gently to the N. Type locality of Takhtsang gneisses, the main crystalline rocks of Bhutan. (See also colour photo of Book Cover)

103). Garnet-sillimanite-biotite schists are also seen within the gneisses. As a whole the characteristic gneisses are widespread all along this northern belt of Bhutan. They outcrop again in the Wang Chu above Thimphu and form near continuous outcrops along the Mo Chu, where, interrupted only by the marble zone of Tamji, they extend northwards to the Tibetan border, culminating in the impressive Masang Kang mountain group (phot. 47). The gneisses along the Mo Chu differ, however, from the Takhtsang type by their less uniform composition, due to granitization, by frequent migmatitic zones and often prolific dyke intrusions associated with the regional young tourmaline granite intrusions along the Tibetan border. The lenticular calc-silicate concretions are less frequent.

In N-central Bhutan, Takhtsang type gneisses are widespread around the Bumthang area. They rise E and NE of the Cändebi carbonate zone and N of the phyllitic rocks already described from the middle Tongsa-Chu. The gneisses are well exposed in the gorge leading to the famous Tongsa Dzong, the birthplace of united Bhutan. For this central part of Bhutan I introduced the local name of Tongsa gneisses, though I associate them with the Takhtsang gneisses of western Bhutan, with which they have many features in common (*Gansser*, 1964). The Tongsa gneisses are characteristically banded and striated (Streifengneiss) and some of the bands expose minute but intense southwards-directed flow folding (fig. 104). They consist of muscovite-biotite-granite gneisses and differ by the larger amount of

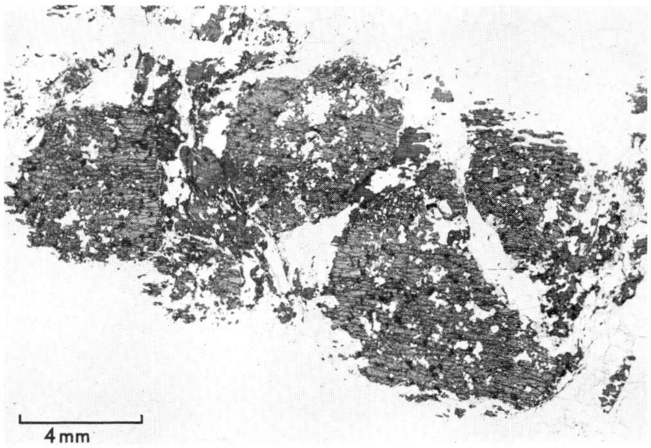

Phot. 45. Microfractured garnet from garnet amphibolite at assumed thrustline between Paro metasediments and Takhtsang gneisses. N of Paro (see also fig. 71).

muscovite and the scarcity of garnets from the otherwise similar Takhtsang gneisses. Migmatitic zones and locally fully granitized and mobilized sections are often met.

In the northern Bumthang region ENE of Tongsa the banded gneisses outcrop again, dipping to the N. They can be followed until the conspicuous intercalation of the steep dipping Djüle La zone which separates them from a more northern granite gneiss zone, where younger granites and their related dyke system are particularly widespread.

From the Lingshi basin to the NE along the Mo Chu, the Takhtsang gneisses extend into the Masang Kang area which we have already mentioned in connection with the northern Tethyan sediments of the Toma La. Along the Mo Chu the Takhtsang crystalline forms a wide basin subdivided by the Tamji marbles. Further north is the culmination of the Masang Kang mountain group (fig. 105). A broad antiform coincides with the Masang Kang summit. Its axis strikes approximately EW and continues

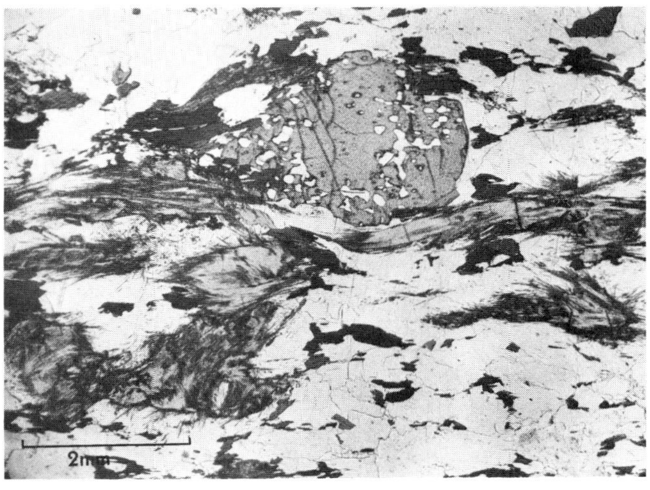

Phot. 46. Garnet-sillimanite-biotite gneisses. The garnets show often quartz inclusions, the sillimanite crystalls have borders of fibrolite. Takhtsang gneiss from Takhtsang type locality.

westwards into another high mountain group, the Kangcheda WNW of Laya (fig. 106, phot. 48).

The basal rock type is still a biotite-(muscovite)-garnet gneiss, often with sillimanite rich zones. In the Masang Kang region these gneisses have been strongly mobilized and locally granitized, as evidenced by a prolific system of dyke intrusions and widespread migmatites (phot. 49). Most of the dykes seem to have some relation with the Chomolhari granite intrusion, but nowhere in this area do tourmaline granites form any substantial mass, in contrast to the Chomolhari and the more eastern region.

On the S side of the Masang Kang dome occur boudinaged garnet amphibolites with lenses of ultramafic rocks and calc-silicate layers. They indicate a tectonic divison between the gneissic horizons. The garnet amphibolites are of special interest, with garnets forming sieve-like inclusions in large hornblende crystals, and each garnet is surrounded by a white rim of andesine-labradorite which seems to have partly replaced the originally larger garnet porphyroblasts (phot. 50). Equally important are lenses of ultramafics. A fresh specimen consisted of a chlorite-spinel-enstatite-pargasite-olivine fels. The olivine (Fo 90) is partly kinkbanded. The brown spinel changed locally into magnetite and chlorite. Other lenses of the same belt are strongly serpentinized. The metamorphism affecting this ultramafic rock is of the sillimanite grade which coincides with the grade of the surrounding formations. The relations of the ultramafic lenses with the garnet amphibolites which belong to the same belt is not clear. The contacts seem to be mostly tectonized.

Within the more migmatitic gneisses occur fully mobilized medium to fine grained granites, which begin to intrude discordantly into the otherwise rather diffuse migmatites. Such locally developed granitic zones are frequent in the highly migmatized main body of the Masang Kang peak (fig. 107; phot. 51). The granitized migmatites form some of the wildest and most spectacular peaks in the Bhutan Himalaya, well displayed in the ranges opposite Masang Kang (phot. 52).

Microscopically, the patchy granites of Masang Kang are seen to consist of large orthoclase, some with myrmekitic reactions and strongly zonar plagioclase, with a border of oligoclase and a core of basic andesine. Striking is the presence of larger cordierite crystals with some type of chloritization along the crystal fractures. Locally, grains of andalusite complicate the paragenesis of these rocks. Quartz is most irregular and often occurs as inclusions in the feldspars or borders them irregularly; it looks as if it were partly introduced. Biotite is generally common, together with some muscovite, but tourmaline is missing. The biotite and muscovite are often intergrown. In other granites only biotite is present. The more basic zoned plagioclases, the quartz infiltration, the predominant biotite, the cordierite and the lack of tourmaline distinguish these older granites from the youngest leucogranite types (see also chemical analysis and ages, p. 132).

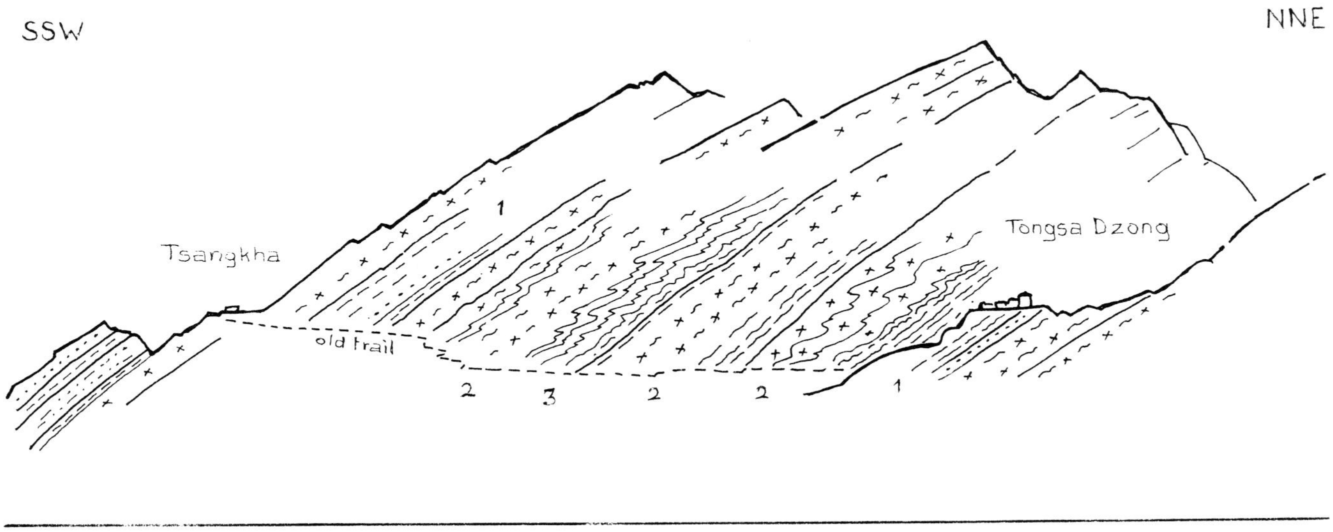

Fig. 104. Tongsa gneisses W of Tongsa.
1 quartzitic mica schists, *2* banded granite gneisses (Streifengneiss), locally folded, *3* muscovite-biotite schists with small drag folds, foldaxis dipping with 8° to SW. Axial planes horizontal. Towards the W increase in pegmatites.

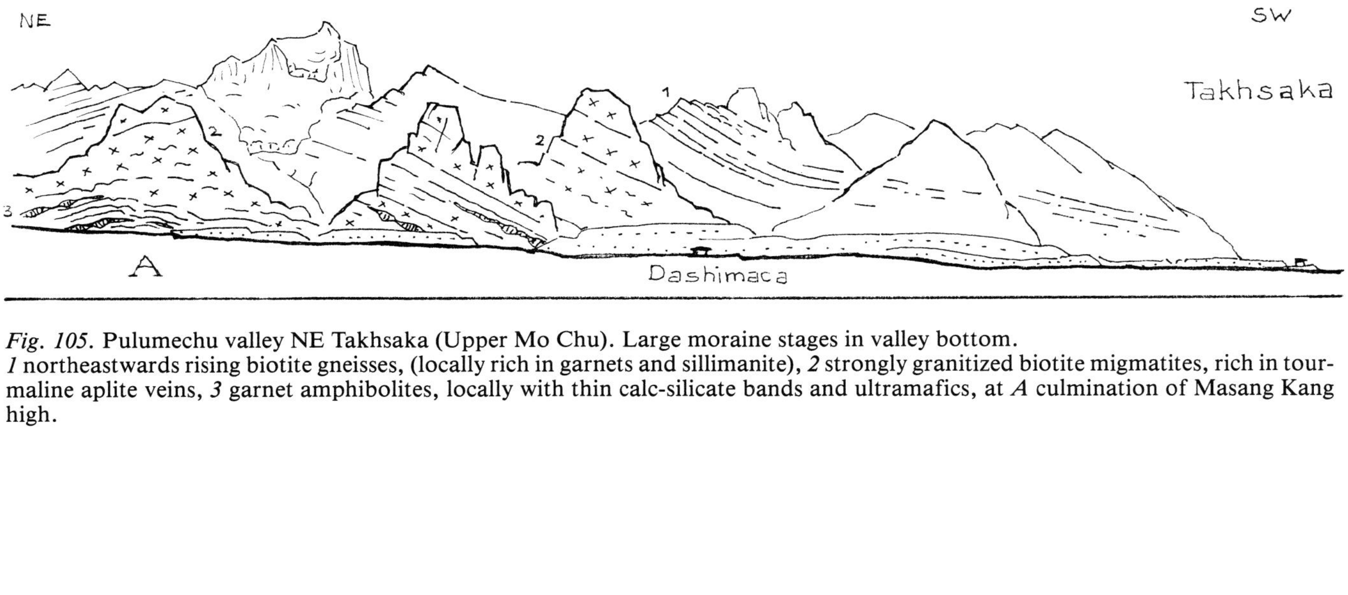

Fig. 105. Pulumechu valley NE Takhsaka (Upper Mo Chu). Large moraine stages in valley bottom.
1 northeastwards rising biotite gneisses, (locally rich in garnets and sillimanite), *2* strongly granitized biotite migmatites, rich in tourmaline aplite veins, *3* garnet amphibolites, locally with thin calc-silicate bands and ultramafics, at *A* culmination of Masang Kang high.

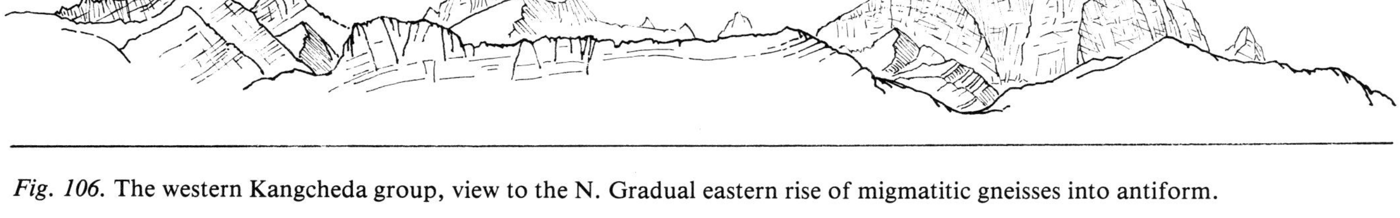

Fig. 106. The western Kangcheda group, view to the N. Gradual eastern rise of migmatitic gneisses into antiform.

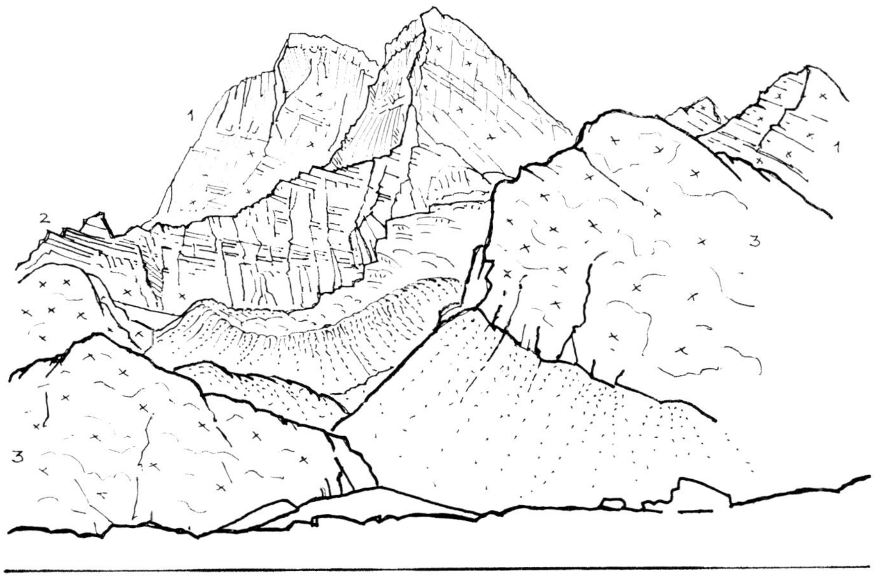

Fig. 107. The double summit of Masang Kang, eastern side. Sharp moraines show rapid glacier movements.
1 upper well bedded migmatitic biotite gneisses: base of Toma La sediments,
2 banded biotite-psammite gneisses and schists with some amphibolites (rusty weathering),
3 lower, more massive and irregular migmatite, more pronounced granitization and aplitic dyke intrusion.

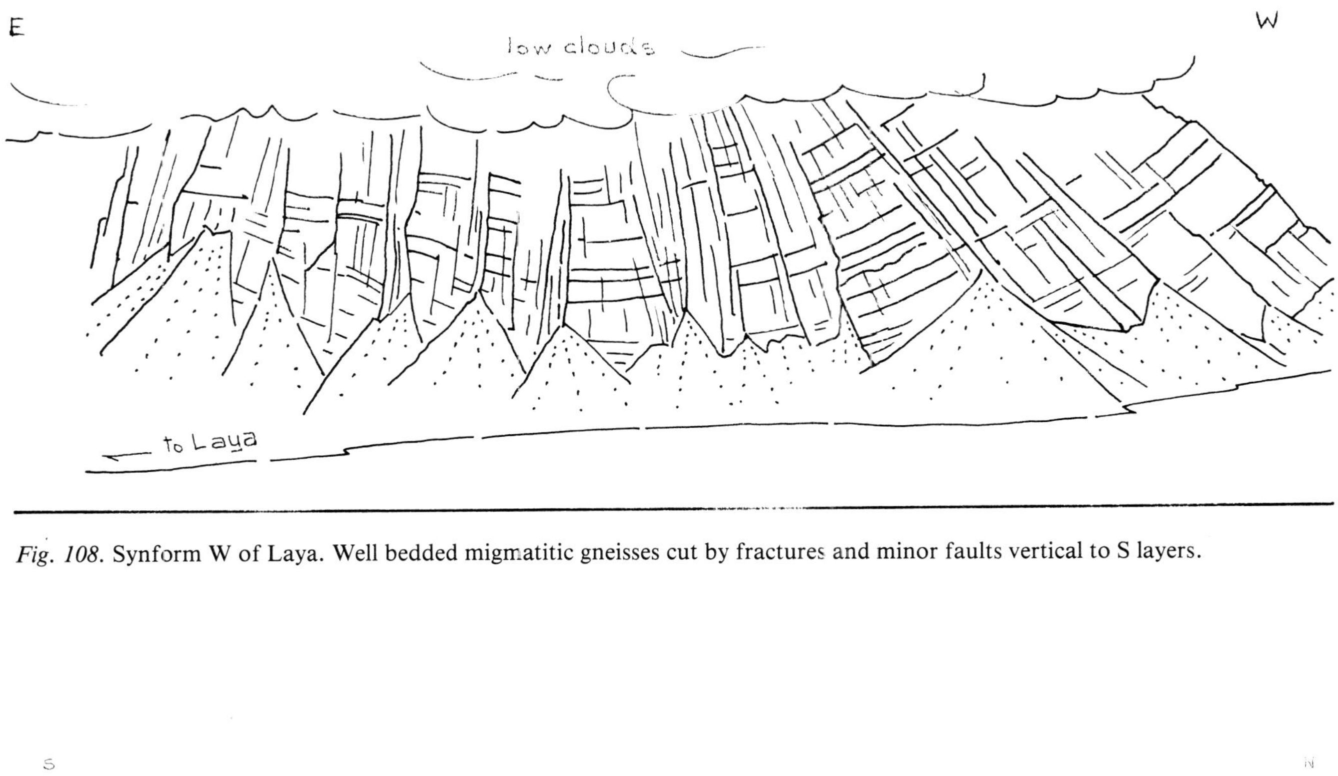

Fig. 108. Synform W of Laya. Well bedded migmatitic gneisses cut by fractures and minor faults vertical to S layers.

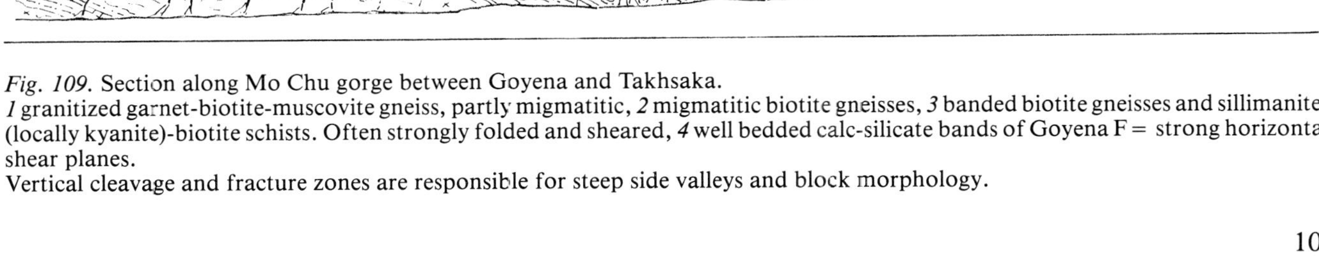

Fig. 109. Section along Mo Chu gorge between Goyena and Takhsaka.
1 granitized garnet-biotite-muscovite gneiss, partly migmatitic, *2* migmatitic biotite gneisses, *3* banded biotite gneisses and sillimanite-(locally kyanite)-biotite schists. Often strongly folded and sheared, *4* well bedded calc-silicate bands of Goyena F = strong horizontal shear planes.
Vertical cleavage and fracture zones are responsible for steep side valleys and block morphology.

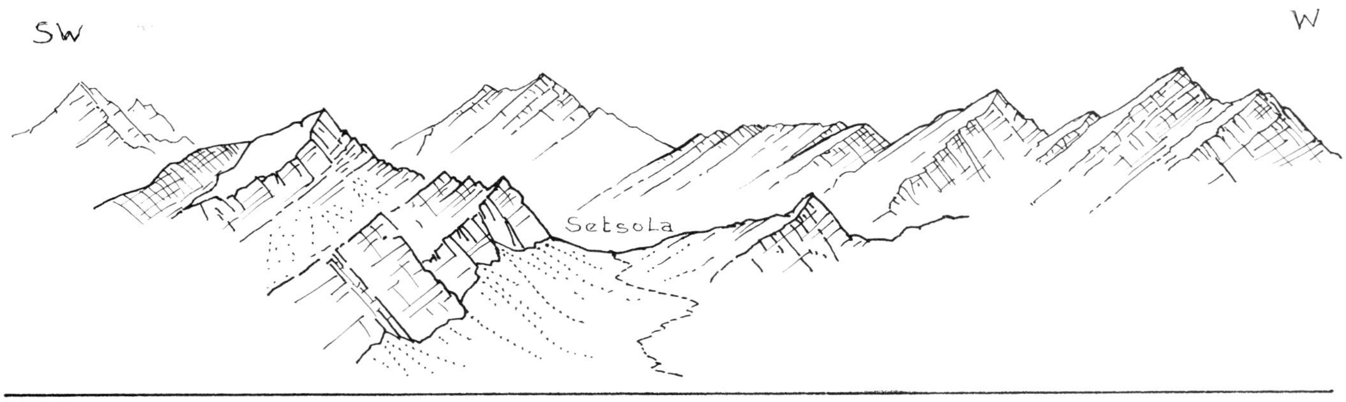

Fig. 110. Setso La (4500 m) with well bedded SE dipping migmatitic gneisses. Seen from Tamgoro.

A prolific dyke system varying from pegmatites, tourmaline granites and aplites crosses the more or less migmatized gneisses (phot. 53). Many of the coarser dykes have a fine aplitic border zone, which sometimes seems to branch into aplitic sill-like bands intercalated with the already migmatitic gneisses (phot. 54). Some of the dykes clearly follow preexisting fault zones within the well banded, often psammitic gneisses, where the rather sharply defined psammitic bands indicate the amount of displacement (phot. 55). One can also note how fine grained aplite granites finger out into highly migmatitic but still banded gneisses (phot. 56).

To the W the Masang Kang high can be followed with all its complex rock formations into another high mountain group, the Kangcheda, W to the village of Laya (fig. 106, phot. 48). The predominant rocks are here garnet-sillimanite-biotite gneisses of the Takhtsang type. They are mostly migmatitic and can grade by an increased mobilization into biotite-(muscovite) granites. Gneisses and intercalated schists expose a distinct lineation and small folds with axis dipping with 10°–30° to the N. The granite is rich in K-feldspar with some microcline crosshatching. The plagioclases show myrmekitic reactions with the K-feldspars. Graphic granite veins, rich in garnet, occur locally. Tourmaline is practically absent and the granites resemble the Masang Kang type rather than the younger tourmaline bearing leucogranites.

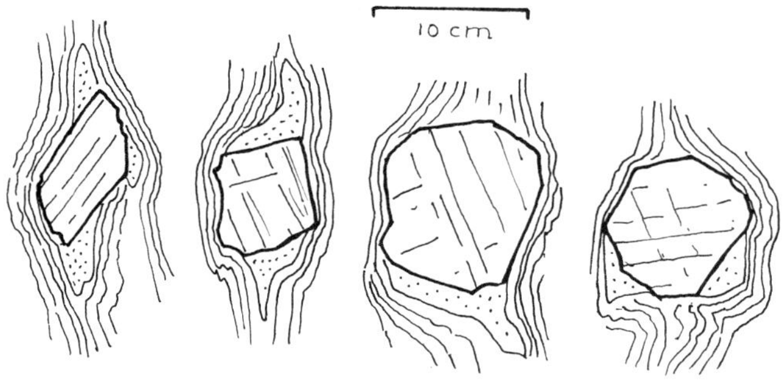

Fig. 111. Large K-feldspar augen with quartz in pressure shadow zones surrounded by fine grained biotite gneiss. Tsephu type augengneiss.

Between Kancheda mountain and the village of Laya the migmatitic and sillimanite bearing biotite gneisses strike NS and dip to the E but further to the E they gradually change and dip NW and W, suggesting a synform. A part of this structure is visible in the steep cliffs along the S side of the valley. Here a very marked fracture system perpendicular to the schistosity of the gneisses (which corresponds to "S") further outlines the synform (fig. 108). N of Laya the migmatitic biotite gneisses are intensely folded and discordantly cut by fully mobilized aplitic granite zones.

From S of Laya, along the Mo Chu until the famous Dzong of Gaza the Masang Kang type crystalline is well exposed along the wild river gorge. The northern part, down to the small place Goyena (N of Gaza) is strongly tectonized with conspicuous near horizontal shear zones (fig. 109). The regionally N dipping gneisses form the southern flank of a synform, the axis of which passes just S of Laya. Migmatized sillimanite and garnet bearing biotite gneisses represent the dominant rock type. Certain parts are more schistose and show strong disharmonic folding. The shear zones however do not follow the more schistose beds but cut discordantly through any rock type. Granitization is locally quite pronounced with the formation of massive biotite granites, however without tourmaline. The latter is frequent in aplitic dykes and fine grained, discordant leucogranite lenses and small stocks, apparently younger than the normal biotite granite type. This complex rock association can be followed southwards to Gaza, where opposite the Dzong, near vertical cliffs are cut by numerous vertical fractures responsible for a peculiar morphology. Here the steepest rock slopes are covered by a luxuriant bamboo jungle (phot. 57). Just below Gaza Dzong, on the southern edge of a landslide, issue several hot springs which since ancient times are famous for their healing power. The 43°C hot water is sulfurous, slightly ferruginous with some CO_2 gas. It has a marked alum taste. It originates probably from the steeply fractured migmatitic gneisses, which further to the S, in

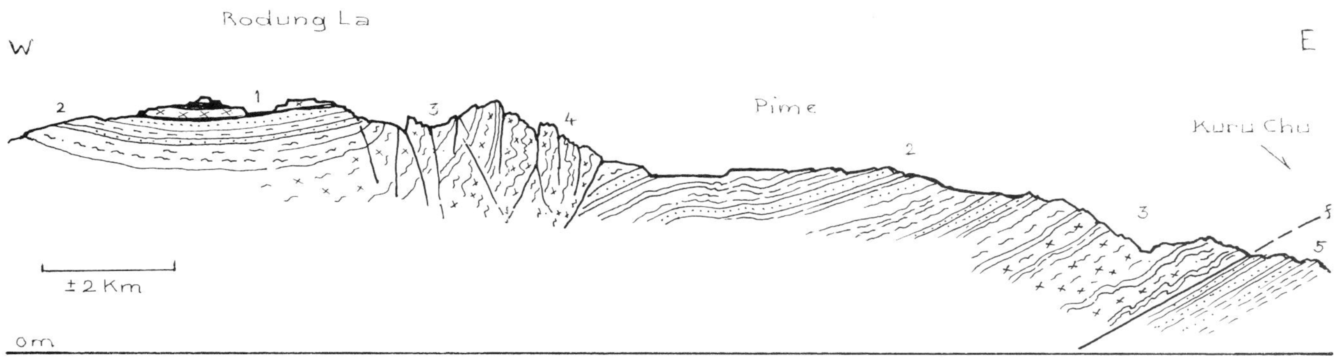

Fig. 112. Eastern flank of Rodung La.
1 leucogranite sill in amphibolites, *2* banded quartzites with muscovite schists, *3* muscovite-biotite streifengneiss, *4* biotite migmatite gneiss, folded with axis 20/15, *5* metasediments of Kuru Chu-Shumar spur below thrust.

the Tamji region, contain intercalated marble zones. Along these marbles we have already mentioned the hot springs from below Tsephu further to the E.

A highly migmatized gneiss area with mobilized younger granites was observed between the Masang Kang region in the N and the Punakha area in the S along the southern trail which runs from the Mo Chu to the western Lunana. This area follows to the E of the Mo Chu section just mentioned above. After crossing the Setso La one reaches the wider Tamgoro area in the upper Rong Chu, a wild tributary to the Pho Chu which drains the southern Lunana region. Here the migmatitic biotite gneisses show a very marked regional dip of 30° to the SE which characterises the morphology of the wider area (fig. 110). At Tamgoro the migmatites alternate with sillimanite-biotite-garnet schists and gneisses and are cut discordantly by fine grained young muscovite-biotite leucogranites, here without the otherwise characteristic tourmaline content. Most likely these leucogranite intrusions continue and increase towards the western Lunana where they are already more widespread along the southern side of the valley.

Further to the SW, on the eastern side of the Mo Chu, steep cliffs consist of a most conspicuous giant augengneiss, well developed at the small rock monastery of Tsephu (phot. 58). K-feldspar porphyroblasts up to 15 cm large form the giant augen which can be idiomorph but also, perfectly rounded. They are rotated with new quartz crystallized in the pressure shadows (fig. 111). It is interesting to note that such augengneisses occur frequently along the marginal part of the large gneiss nappes.

The southern border of the Masang Kang type gneisses, migmatites and granites reaches the Thimphu area where I have already described the gradual contact towards the Paro metasediments (fig. 72). We have also noted how Takhtsang type gneisses rise to the S below the thick basal marble zones of the southern Lingshi basin. Here, along the Wang Chu , N of Thimphu, outcrop migmatitic garnet bearing sillimanite-biotite gneisses of the Takhtsang type. They form the rock walls of the famous Cäri Gömpa. In the neosome of the more migmatitic parts large K-feldspars are frequent, always with the conspicuous rim of biotite. The northwards dipping gneisses are locally folded with fold axes plunging towards the NW.

Well-known are the Thimphu gneisses, since they have been quarried NW of the capital for the reconstruction of the large Dzong. These migmatitic gneisses form the very base of the northern crystallines. They have been studied in detail by R.*Hänny*. The Thimphu migmatites are strongly lineated. In sections parallel to the main lineation the leucosome appears as irregular parallel bands of a few cm thickness and up to 1 m in length (phot. 59). Perpendicular to this lineation the veins are tightly folded in a «ptygmatic» style with cm amplitudes and with a rather constant orientation of the axial planes (phot. 60). The leucosome is bordered by a few mm thick rim of coarse grained biotite. The paleosome of the migmatites is fine to medium grained and consists mainly of quartz, K-feldspar and plagioclase, some garnet and biotite. Here, garnet is mostly fine grained and occurs often with large K-feldspar, showing an irregular microcline pattern, plagioclase and quartz. Abundant myrmekite lobes protrude into the K-feldspar. Sillimanite was not observed in the Thimphu migmatitic gneisses but appears further to the WNW above Pachodin Gömpa. Here in the High plateau with the beautiful glacial lakes we note a N dipping section of biotite migmatites, biotite-garnet-psammite gneisses and garnet-sillimanite-biotite gneisses. Intercalated are bands of garnet amphibolites with strongly poikiloblastic garnets where the inclusions consist of quartz and plagioclase though hornblende is missing. Also intercalated are biotite-pyroxene-hornblende gneisses with large fresh plagioclase (labrador).

From the Masang Kang area to the E, the northern crystalline belt has been investigated in the wider Lunana region, in northern Bumthang, in the Kuru Chu, Koma Chu and the Yangtse Chu above Yangtse Dzong. In all these areas I have already described the intercalated metasediments.

The actual Masang Kang antiform can be traced into the wild border range of the northern Lunana region. The top of the high range dips to the N and is

Phot. 47. The Masang Kang group (7200 m) W side. Gentle northern dip of migmatitic gneisses with biotite granite bodies. Masang Kang antiform just to the right.

covered by the lowest Tethyan sediments, mentioned already and shown in fig. 50. The main crystalline body, forming the base of the sediments, resembles very much the rock types of the Masang Kang area. Migmatites dominate, here with a higher percentage of the neosome as well as the fully mobilized granitic part. The migmatites are often broken up into irregular «Schollen» within the granite intrusions (phot. 61, 62, 63). The muscovite-biotite granites are finer grained than in the Masang Kang area but still without tourmaline. However, they appear to belong already to the widespread younger leucogranite type, which is also characterized by a higher muscovite-biotite ratio and the generally larger muscovite grains. In the eastern Lunana, towards the Gonto La, larger bodies of leucogranite and surrounding migmatites are here highly fractured, a fact otherwise rare in the stronger granitized zones. Two main fractures cut the granites: the main fracture dipping with 80° to the NW and a secondary fracture dipping with 25° to the ENE. The extension of this fracture-zone is not visible since the area is covered by large glaciers, but the amount of morainic material is here particularly rich suggesting a wider extent of this tectonized zone (phot. 64). In the Kephu region (western Lunana) younger quartz veins contain pyrite and

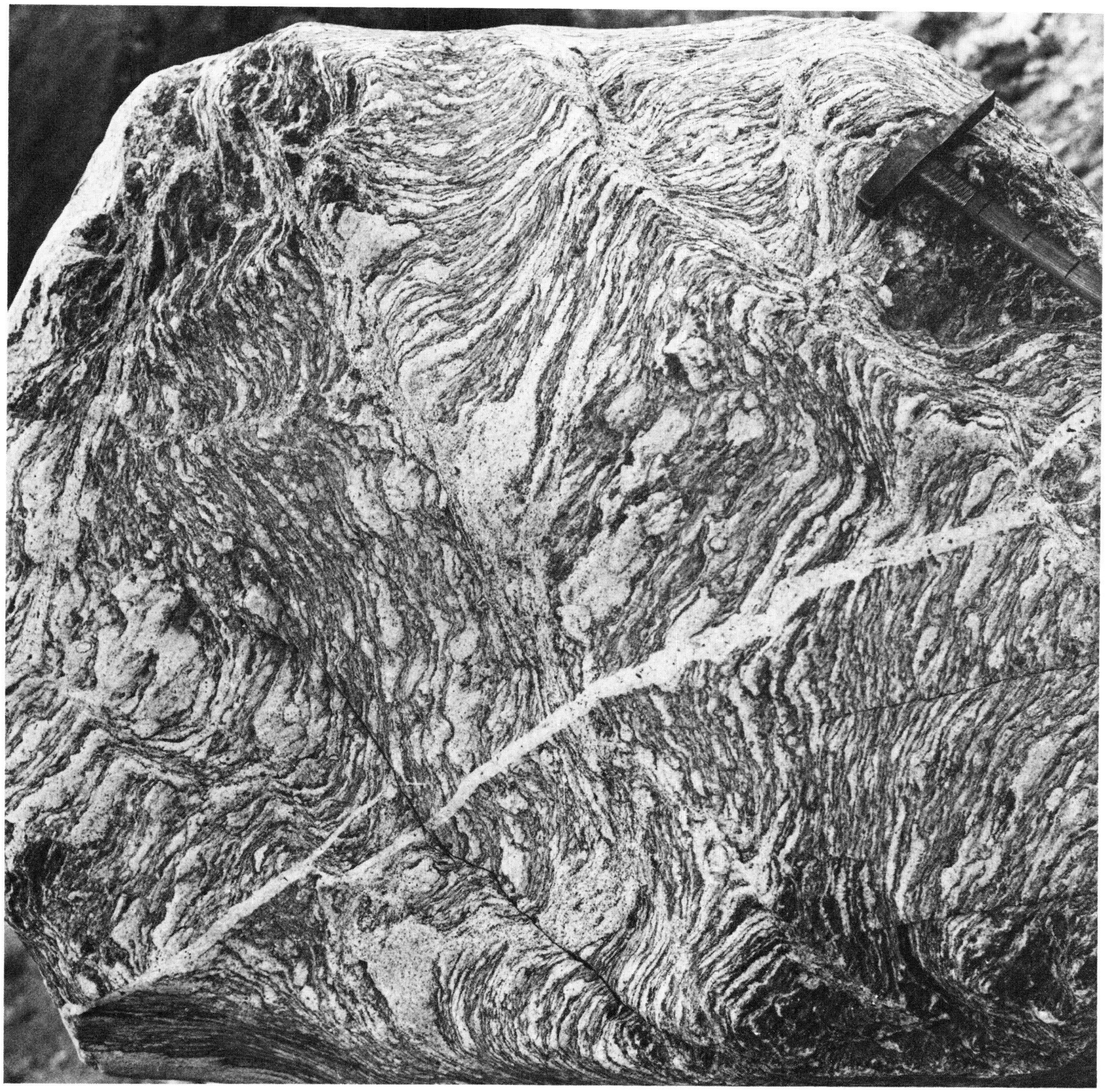

Phot. 49. Typical advanced migmatite of Masang Kang type, cut by thin tourmaline aplite. Locally granitized patches. S Masang Kang.

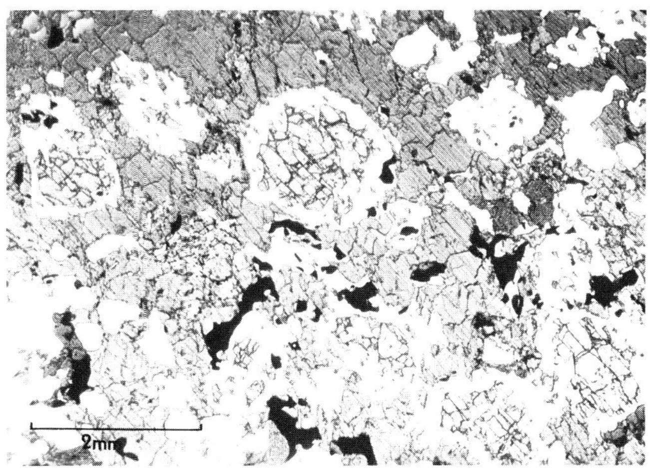

Phot. 50. Garnet amphibolite. The garnets grow in large hornblende crystals and show reaction rims of andesine-labradorite which seem to have replaced the original garnet. S of Masang Kang.

Phot. 48. The Kangcheda group (6800 m) SE face. Horizontal migmatitic gneisses indicating the crest of the Masang Kang antiform.

Phot. 52. The wild range of granitized migmatites SE of Masang Kang ending in the over 7000 m high Tsenda Kang (left side). Seen from NNE. The gently folded gneisses are cut by steep, NS directed fractures and faults, responsible for the rough topography.

Phot. 51. Masang Kang main peak with granitized bodies within migmatites and biotite-sillimanite gneisses. Some of the local biotite granites contain cordierite. Looking at steep SE face.

galena. The mineralized zones are rusty weathered and can be easily detected in the field. They postdate the granitization.

An excellent section of the northern crystalline is exposed along the Chamka Chu between Tsamba in the NW and Kakhtang in the SE (fig. 75). Both ends of this section have already been mentioned in connection with the overlying and underlying metasediments. The Chamka Chu forms a most impressive gorge freshly cut into a wider and older valley torso. Below the cover of the Tsamba marbles follow banded biotite-(muscovite) gneisses of the «Streifengneiss» type. In the central part they become migmatitic and locally granitized without however discordant contacts of the biotite granites which resemble the Masang Kang type. The basal part of the cliffs often expose more thin bedded gneisses which are intensely subfolded with EW striking fold axes which dip gently to the W or are horizontal. Axial planes are mostly steep. A prolific intrusion of muscovite-pegmatite dykes is often concentrated in those sections where the folding is most intense. A certain relation of folding and intrusions exists, though the younger dykes are clearly cutting the folds. In general the gneisses are flat lying and cut by vertical fractures which are responsible for the

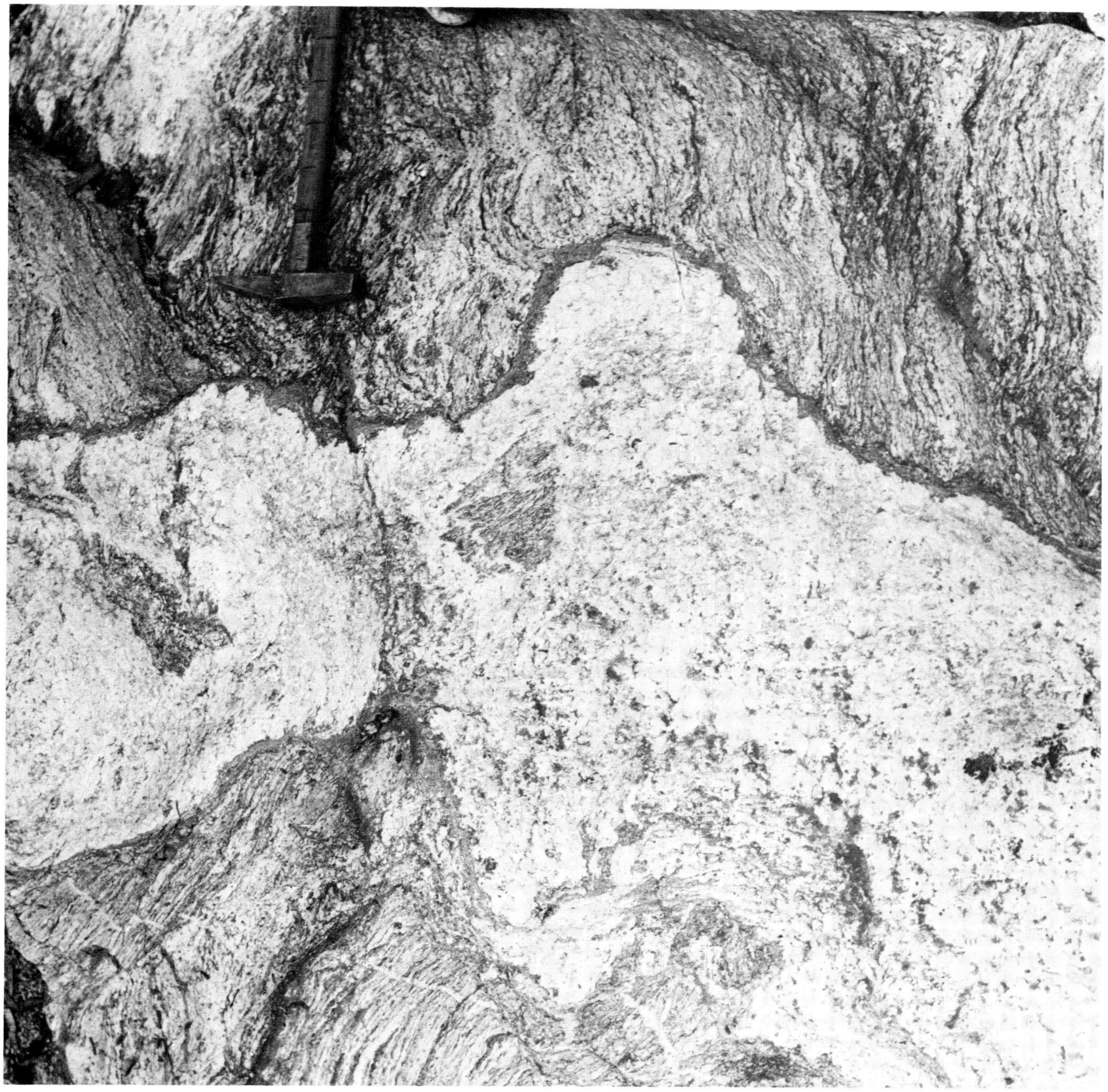

Phot. 53. Pegmatitic muscovite granit dyke intruding migmatites S of Masang Kang. Fine biotite is concentrated along border which can become locally more diffuse (lower right side). Note xenoliths of migmatite.

spectacular rocktowers exposed on both sides of the gorge. Southwards the flat gneiss layers rise gradually to the SE and at the same time biotite-psammite gneisses replace the migmatitic gneisses and the granitized parts. We have already noted how the psammite gneisses are eventually thrust without however a clear cut thrustline over the staurolite schists belonging to the Bumthang metasediments (fig. 75).

Between Bumthang and Lhüntse Dzong in the Kuru Chu valley, the southern part of the northern crystalline zone was followed along the trail over the Rodung La. The top of the pass exposes flat lying, well bedded garnet bearing staurolite-biotite-psammite gneisses alternating with quartzitic muscovite schists and some banded amphibolites. They are intruded by fine grained tourmaline granite sills and the whole section resembles a basal part of metasediments topping the main crystalline sheet. Eastwards the banded sequence rises to the E and underneath appear striated and folded biotite gneisses grading into the typical garnet bearing biotite migmatite gneisses (fig. 112). Locally the western dip steepens to 80° while the fold axes plunge gently to the NNE. This part of the section is criss-crossed by various fracture systems. Further to the E,

Phot. 54. Aplite-granite dyke intruding banded, migmatized, biotite-psammite gneisses. Locally small apophyses of the aplitic border enter the country rock. S side of Masang Kang.

Phot. 55. Granite aplite dykes following some faults while intruding banded migmatized psammite gneisses. Note the displacement of the fine grained biotite psammite gneiss layer. S of Masang Kang.

Phot. 56. Aplite granite sill fingering out into banded migmatite gneiss. (Aplite 40 cm thick). S of Masang Kang.

56

55

Phot. 57. The Mo Chu valley N of Gaza with dense bamboo jungle covering steep slopes of migmatitic biotite gneiss with local granitization. The slopes are cut by a steep fault and fracture system. Looking towards SE.

Phot. 58. The Tsephu augen gneiss. The K-feldspar porphyroblasts are very irregular from idiomorph to rounded and up to 15 cm large. They form a marginal belt of the main crystalline thrust sheet (see also fig. 111).

Phot. 59. The Thimphu migmatitic gneisses. In section parallel to the main lineation, the neosome appears as parallel bands, a few cm thick and up to 1 m long. Quarry N of Thimphu.

Phot. 60. Thimphu gneisses. Section perpendicular to the main lineation. The neosom, arranged in rod like bands, appears here as irregular knobs, some subfolded. The leucosome patches are bordered by a coarse grained biotite. Quarry N of Thimphu.

Phot. 61. Irregular biotite migmatite, locally granitized. Some aplitic bands contain diopside (below knife). Middle Lunana.

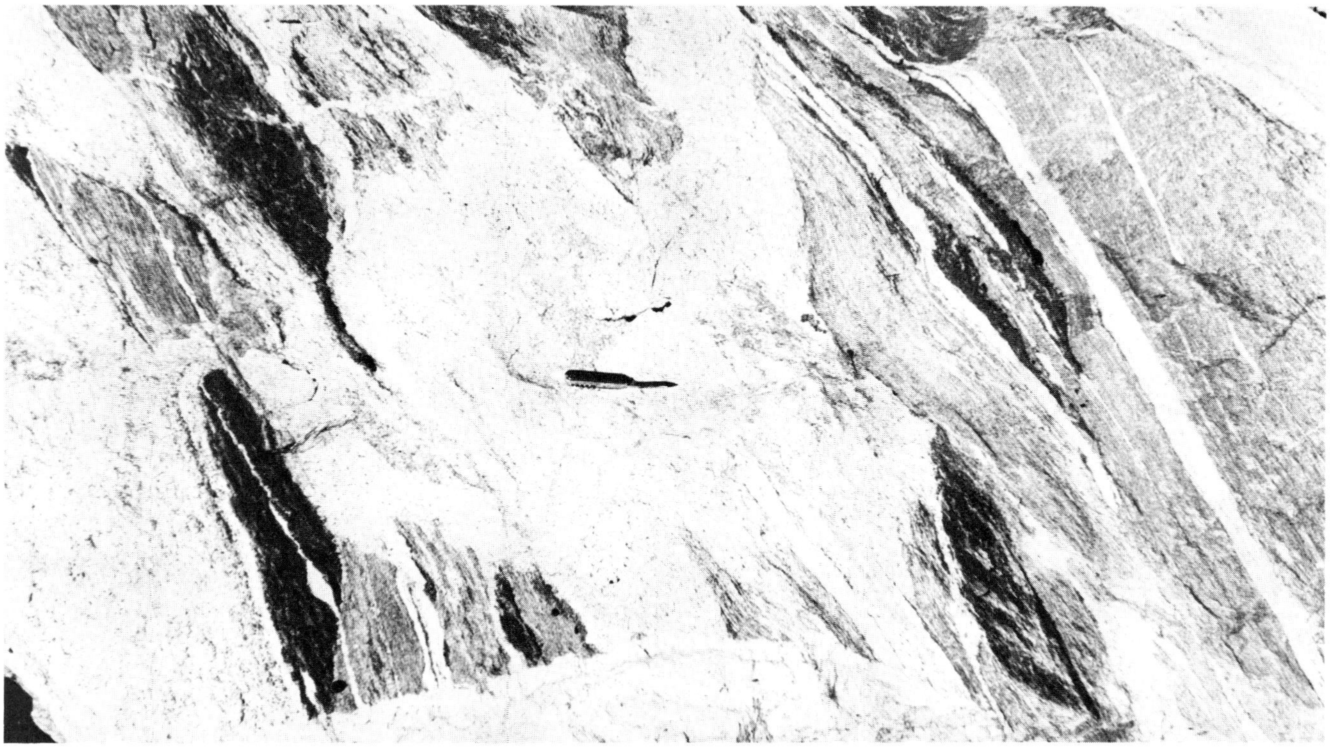

Phot. 62. Biotite – muscovite granite intrudes banded biotite gneiss. Lower part cut by pegmatite with gost-like outline of xenoliths (lower left). Migmatite zone of middle Lunana.

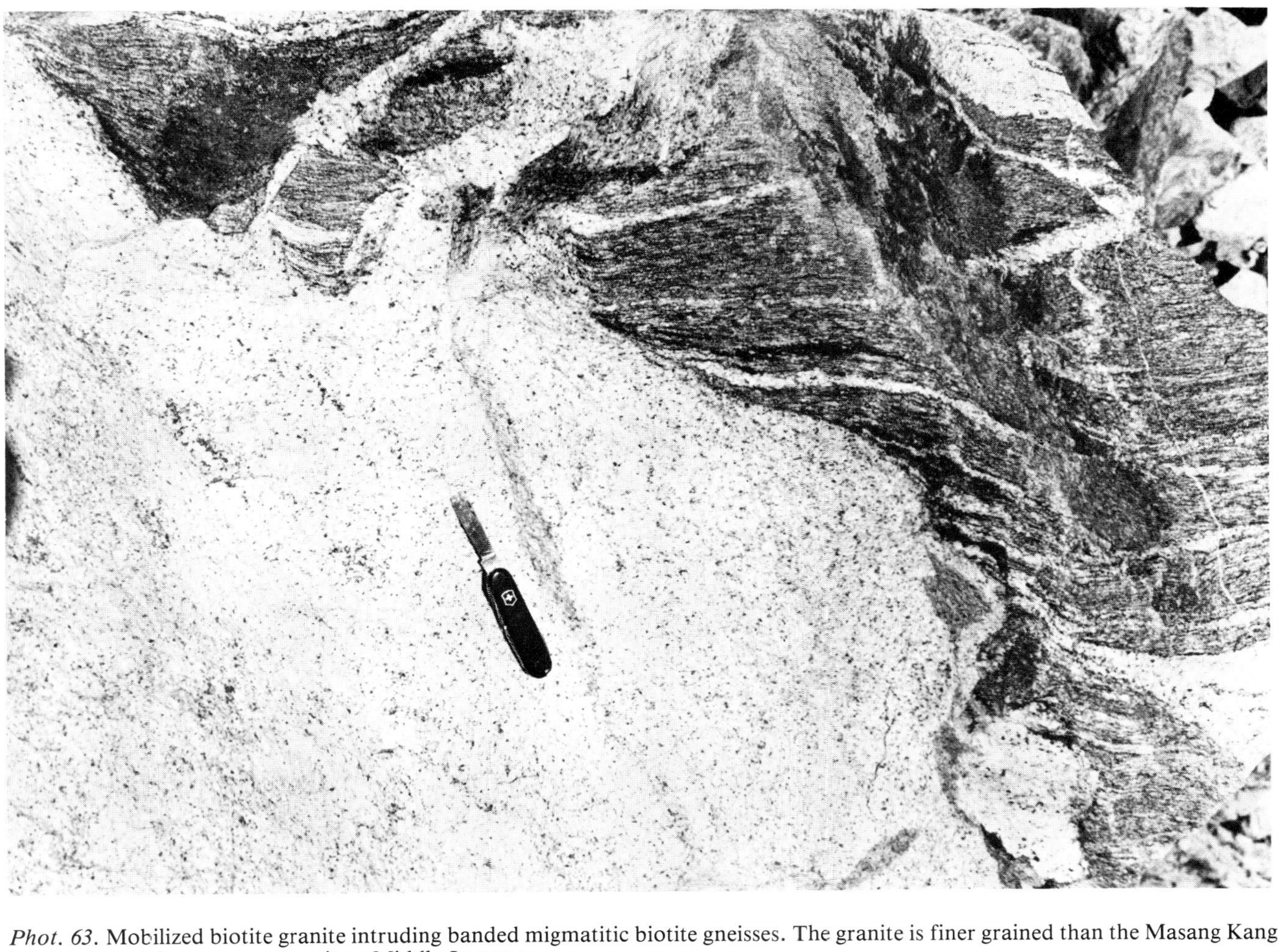

Phot. 63. Mobilized biotite granite intruding banded migmatitic biotite gneisses. The granite is finer grained than the Masang Kang type and contains some muscovites. Middle Lunana.

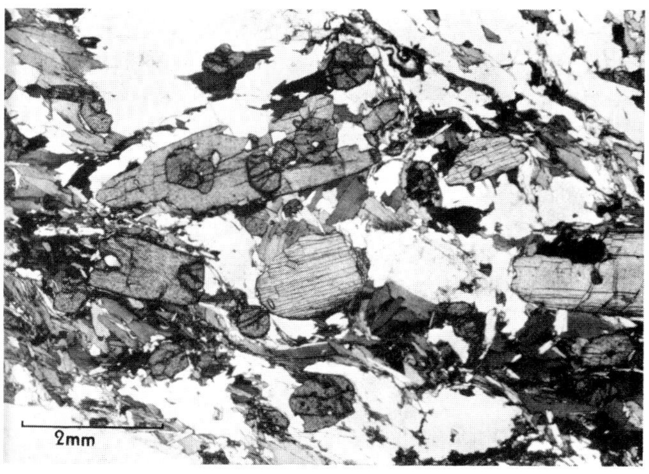

Phot. 64. Steep cleavage and fracturing in leucogranite just E of Monlakarchung La. Same type of tectonisation as near Gonto La. In background biotite gneisses intruded by leucogranites. View towards E.

Phot. 65. Garnet-kyanite-muscovite-biotite schists. The garnets are often included within the kyanite crystals. Southern part of northern crystalline zone E of Rodung La.

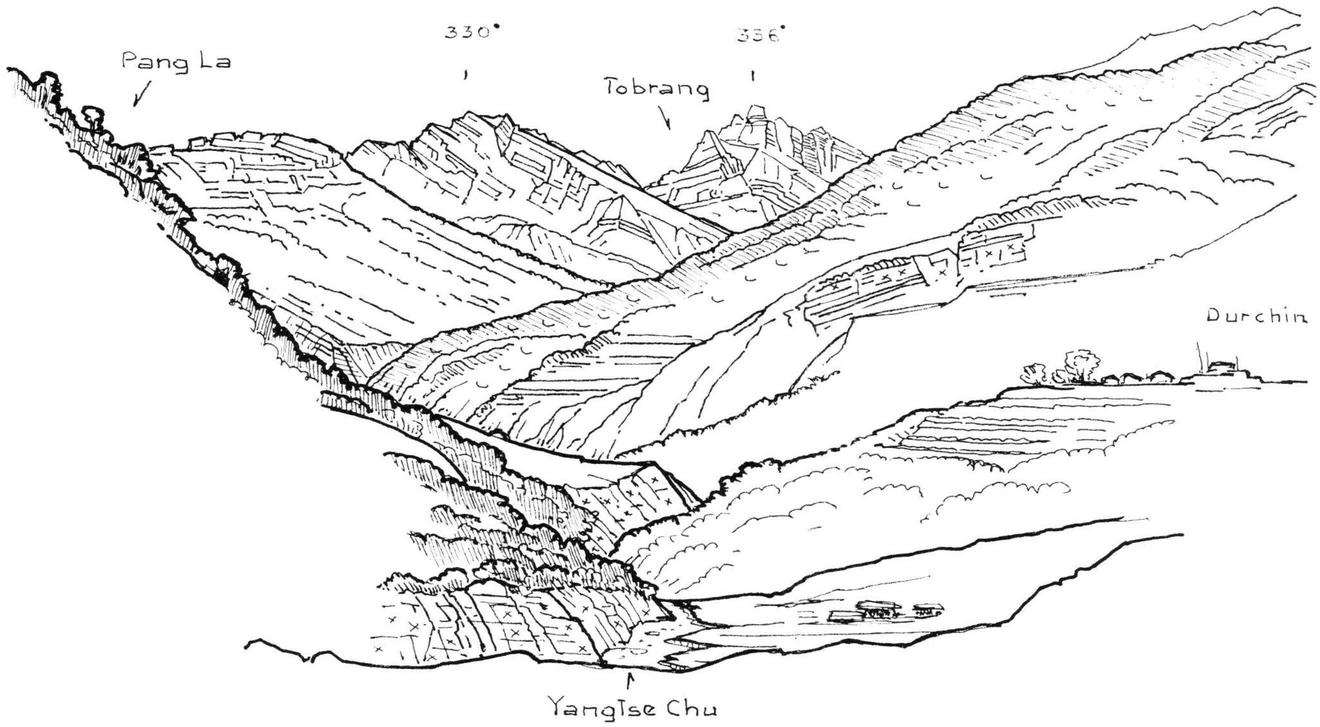

Fig. 113. The upper Yangtse valley N of Chörten Kora. Marked eastern to southeastern regional dip in banded biotite gneisses with intercalated biotite schists and some quartzites. Locally irregular intrusions of migmatite granites. Along the eastern valley slope occur large land-slides. At the bottom northwestwards dipping biotite augengneisses.

now with a west dip of about 25°, appear again well bedded quartzitic muscovite schists, some with a marked lineation and crenulation folds directed towards the SSE. This well bedded intercalation is followed by biotite gneisses alternating with garnet-kyanite-muscovite-biotite schists. The garnets are often included within the kyanite crystals (phot. 65). Eastwards and below reappear garnet biotite migmatite gneisses with an incipient granitization. This rock sequence can be followed until the Kuru Chu where it rises eastwards on top of the N plunging Kuru Chu - Shumar spur.

On the E side of the Kuru Chu spur the Dong La connects the Kuru Chu valley with the Yangtse Dzong in the Yangtse Chu valley. On the W side of the pass pelitic metasediments are widespread such as

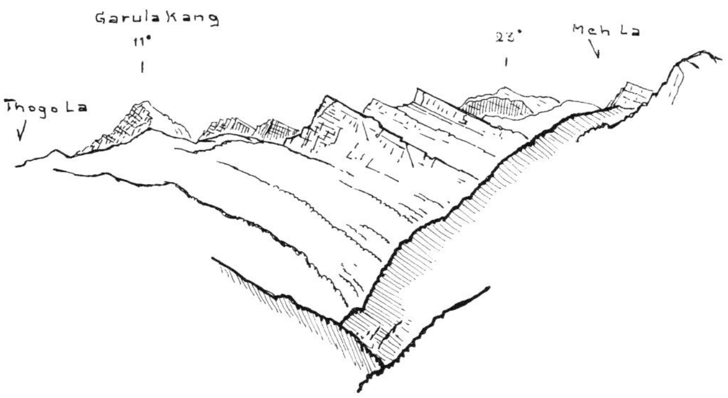

Fig. 114. The head water N of Tobrang. Strongly SE dipping gneisses topped by black biotite schists.

garnet-kyanite-muscovite-biotite schists and schistose muscovite quartzites. Locally some thin calc-silicate bands are intercalated in the more quartzitic horizons. On the pass, banded biotite gneisses of the «Streifengneiss» type set in, together with bands of muscovite quartzites and muscovite schists. They all dip to the E and are locally folded with the fold axis dipping to the N and NNE. Further to the E, biotite augengneisses alternate with coarse grained biotite-(muscovite) flaser gneiss, the latter continuing to the Yangtse Dzong. They also form the outcrops at the famous and largest Chörten of Bhutan, the Chörten Kora, situated to the N of Yangtse Dzong. Sillimanite bearing rocks were not observed along this traverse and migmatites and granitized rocks are absent. The regional eastern dip has gradually changed into a NW and W dip. Further to the N along the Yangtse Chu valley we cross a marked synform after which E dips set in again and dominate the whole northern area (fig. 113). Biotite-migmatite gneisses outcrop in the synform while quartzitic gneisses form frequent intercalations further to the N and are responsible for the excellent bedding in the ranges between the Yangtse Chu and the Koma Chu (fig. 114). The crystalline rocks in the Koma Chu, where several carbonate belts cross the valley, show a somewhat higher metamorphism than in the region further to the E. Thick bedded garnet bearing sillimanite-biotite-migmatite gneisses recall the Takhtsang type and form steep rock walls in the middle part of the valley between the two carbonate zones.

F.6. The Leucogranites of the Bhutan Himalaya

While discussing the northern crystalline units we have frequently met sills, dykes and larger bodies of the mostly tourmaline bearing leucogranites. Identical intrusions are known from various regions, all along the Himalayan range. Apart from widespread small occurrences, mostly as dykes and sills, larger intrusions appear concentrated in certain areas. They are known only in the higher Himalaya within and at the northern edge of the crystalline units and appear frequently related to high and highest elevations: Nanga Parbat in the W, Badrinath and Kedernath in the Kumaon Himalaya, Api and Mustang in NW Nepal, Manaslu and its continuation into Sisha Pangma in northern Nepal and southern Tibet, Everest and Makalu region and Chomolhari and Kankarpünzum region in northern Bhutan. In five areas the granites reach 8000 m and in the rest they surpass the 7000 m elevation (*Gansser*, 1964, 1980; *Le Fort*, 1973), (phot. 66).

It has already been stated that the carbonate zones are frequently intruded by leucogranites, actually most of the leucogranite bodies are in contact with calc-silicate and marble zones. This is not a special case for the Bhutan Himalaya and similar relations have been noted in many sections through the crystalline above the Main Central Thrust. The best example was observed in the Kali river section along the W border of Nepal and along the Pindari glacier (*Heim* and *Gansser*, 1939; *Gansser*, 1964). The same observations can be made on the S side of the Annapurna range in northern Nepal.

In the Bhutan Himalaya the leucogranites are restricted to the northern half of the main crystalline thrust sheets and increase from S to N. They are very rare in the Chasilaka and Sure gneiss sheets, whereas the Tashigang crystalline is intruded by few aplitic and pegmatitic dykes and sills. The general distribution of the granites is shown in figure 125 and 127. We recognize three main centres, all following the northern border area: The Chung La granite in the NW with its extension into the Chomolhari range, the Gophu La granite in the eastern Lunana (northern area) and the Monlakarchung-Pasalum granite E of Kankarpünzum mountain (northeastern area). All three granite bodies are surrounded by a halo of an intricate system of smaller stocks, lenses, apophyses, dykes and sills. Sometimes only the halo is exposed and suggests a larger granite development at depth. However, some larger granites occur in the form of laccoliths and do not increase in size downwards. A good example is exposed in the southern range of Lunana where the higher summits consist of a thick layer of leucogranite which does not reach the valley

Phot. 66. Kankarpünzum, 7550 m. West face with western summit to the right and Chumhari Kang on left side. Dykes and smaller stocks of leucogranite within biotite-sillimanite gneisses and some marble and calc-silicate bands. The leucogranite intrusions increase towards the E.

bottom. Very similar conditions are known from the Manaslu leucogranite (N Nepal) and its eastern extension towards Sisha Pangma (S Tibet) (*Le Fort*, 1973).

The general mode of intrusion, age, petrological and geochemical composition of the leucogranites are surprisingly constant (see chemical analyses p. 132). These features are not only characteristic of the Bhutan Himalaya but indeed all along the Himalayan range (*Le Fort*, 1973; *Gansser*, 1964). The leucogranite of the Bhutan Himalaya is generally medium to fine grained, very homogeneous and mostly massive. It strikes by its light colour with muscovite dominating over the biotite and frequently contains black tourmaline (Schorl). The larger bodies are usually surrounded by an aureol of dykes and sills which can surpass the volume of the central bodies. Xenoliths are practically absent in the main homogeneous bodies, but can be frequent in the more marginal parts. They only represent the englobing rock types (exogene xenoliths) while xenoliths from deeper sources are rare (endogene xenoliths).

Under the microscope we recognize irregular, lobate, larger quartz grains with or without a weak undulatory extinction (30–34%). The K-feldspars have irregular boundaries with the plagioclase in which they can occur as inclusions. They usually show an irregular microcline cross hatching or a perthitic intergrowth (23–30%). The large plagioclases (An 8–12) are generally weakly zoned with dusty inclusions in the more basic core while the rims are clear (32–37%). Myrmekitic reactions with the K-feldspars are frequent. Locally a graphic granite intergrowth can be observed between quartz, K-feldspar and plagioclase. Muscovite often forms larger, well crystallized flakes and is intergrown with biotite, the latter often occurring as central inclusions (10%). The biotite content is very variable but rarely exceeds the muscovite. It has strong chestnut-brown pleochroism (1–3%). Alteration to chlorite is noted in some rocks, together with the formation of ilmenite. Tourmaline is nearly always present but its distribution can be very irregular. Often one notes a roundish concentration (phot. 67) or net like crystals which can form star like aggregates (phot. 68). The macroscopic black tourmaline is mostly zoned, bluish-grey in the centre with olive-brown rims (3%). Rounded grains of apatite are very irregularly distributed. Zircon forms often dichroitic halos in the biotites. Garnet is not frequent except in certain aplitic border zones and dykes.

A few special localities are discussed in the following: In the headwaters of the Paro Chu in NW Bhutan we mentioned the granite intrusion of the Chung La (fig. 23, 115). This leucogranite intrudes the Tremo La phyllites and calc-schists in the Chekha region. The regional contact of the laccolithic body is

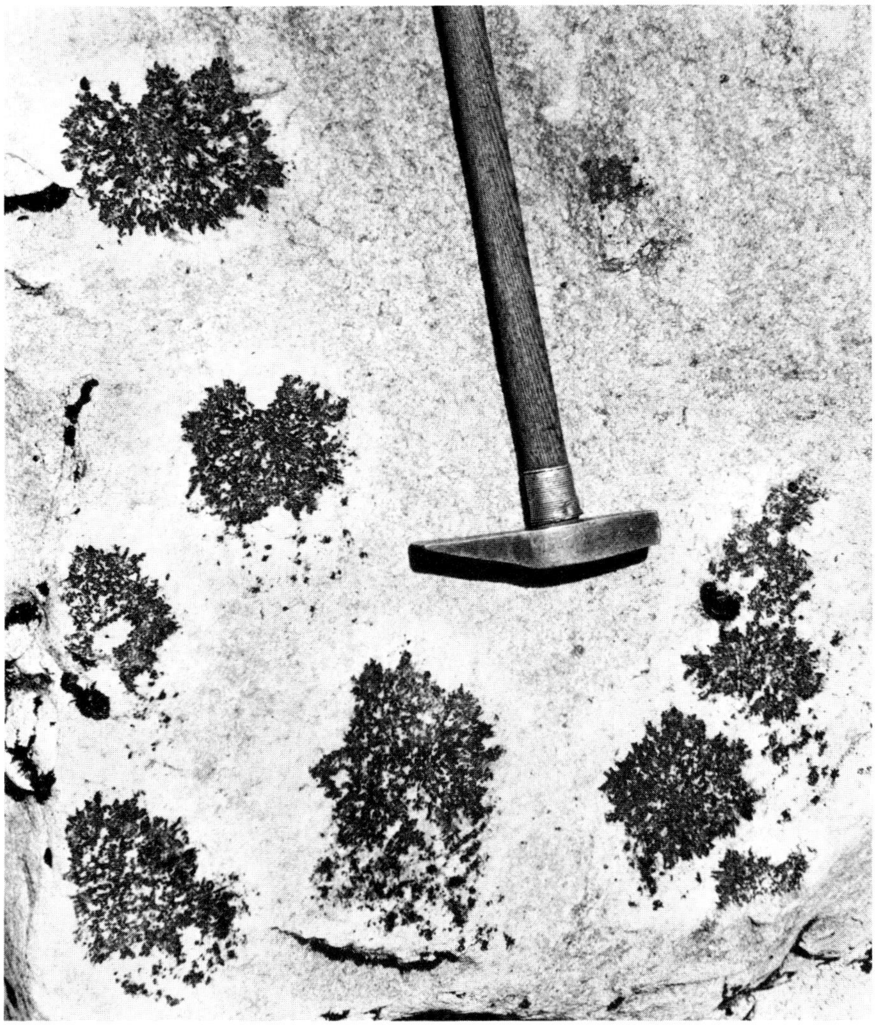

Phot. 67. Aplitic leucogranite with roundish concentrations of tourmaline. From leucogranite sills in the eastern Lunana region.

Phot. 68. Leucogranite with star like arranged tourmaline crystals. Melunghi Chu region, near contact with marbles. The star mesures 5 cm.

Fig. 115. Composite section along upper Paro Chu valley from Chugül to Chekha. (Higher peaks covered by low clouds).
1 sillimanite-garnet-biotite gneiss, *2* granite gneisses to augengneisses, *3* quartzitic intercalations in biotite gneisses, *4* tourmaline granites, partly thick bedded, partly massive, northwards and upwards more aplitic, *5* carbonaceous, fine grained biotite schists (Chekha type).

parallel to the bedding but in detail it is locally discordant (fig. 21). The predominant rock is a fine to medium grained tourmaline-muscovite leucogranite. The quartz has very irregular borders and seems a late crystallization product. Some more aplitic zones show a marked graphic granite texture. Below Chekha the granite is locally fractured with cataclastic quartzes which show a marked, fine grained mortar rim. Thick sills of the Chung La granite extend to the SW into the uppermost Ha valley. They seem to reappear in the ranges of NE Sikkim where they form extensive sills within psammite gneisses and calc-silicate layers well exposed in the S slopes of the Pauhunri-Chomoyummo mountains. While discussing the Tremo La formation we have already drawn attention to this surprising replica of the Chomolhari range NE of the Chung La (fig. 116 in *Gansser*, 1964, and *Auden*, 1935).

The S slopes of the Chomolhari mountain range and its continuation towards Tremo La (fig. 116) are a classic example of leucogranite sill intrusions into metasediments and gneisses. These sills extend from the Chung La granite centre to the NE and gradually die out in the region of Lingshi, a locality already discussed. In the Chomolhari S wall the leucogranite sills intrude a sequence of well bedded more or less migmatitic sillimanite-garnet-biotite gneisses (base) with the already known biotite psammite gneisses, calc-silicate bands, calc-schists and marbles. The whole section dips gently towards the SE where it is cut by the large Lingshi fault which borders the younger sediments of the Lingshi basin. This well bedded pile of parallel rock types is most probably not a normal sequence but is repeated due to isoclinal folding, which is actually visible on a mega scala in the steep cliffs of the Chum Kang, a peak in the northeastern continuation of Chomolhari mountain (fig. 117) and which is also frequently seen with smaller amplitudes (several m) in the rocks of the lower part of the Chomolhari S face (phot. 69). This fact explains the presence of biotite gneisses higher in the section and above some of the larger marble horizons, beautifully displayed in the spectacular outcrops of the Tshering Kang (fig. 118, phot. 70).

The relations of the widespread leucogranite sills with the surrouding carbonate and gneissic rocks are of special interest. The sills are remarkable for their strikingly uniform composition and size. Irrespective

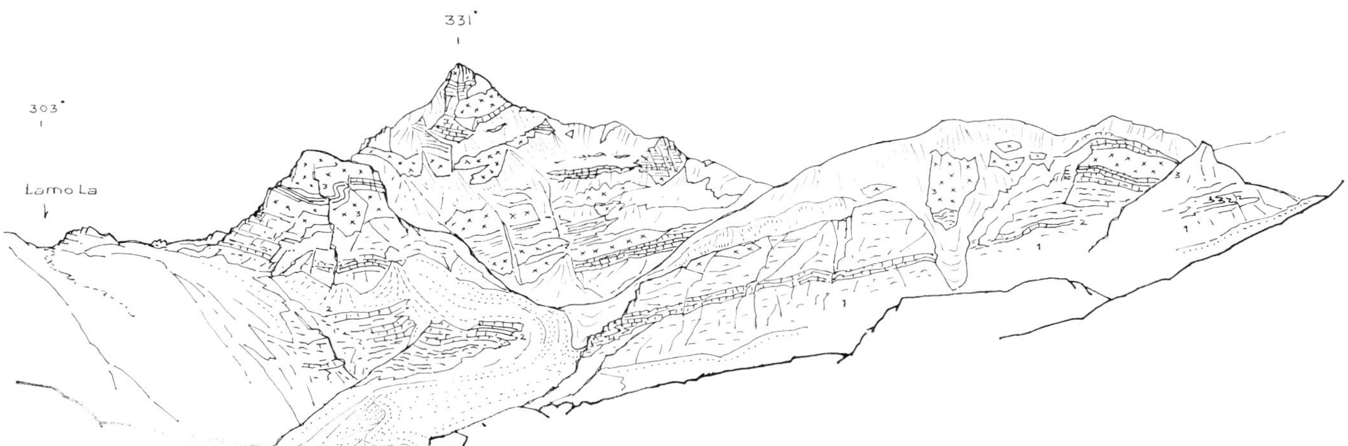

Fig. 116. The border ridge between Chomolhari and Tremo La. Unnamed peak with complicated intrusions of tourmaline granite into biotite-psammite gneiss and calc-silicate bands. Seen from Pt. 4900 m.
1 biotite-psammite gneiss and biotite schists, *2* calc-silicate and marble layers, *3* tourmaline granite intrusions.

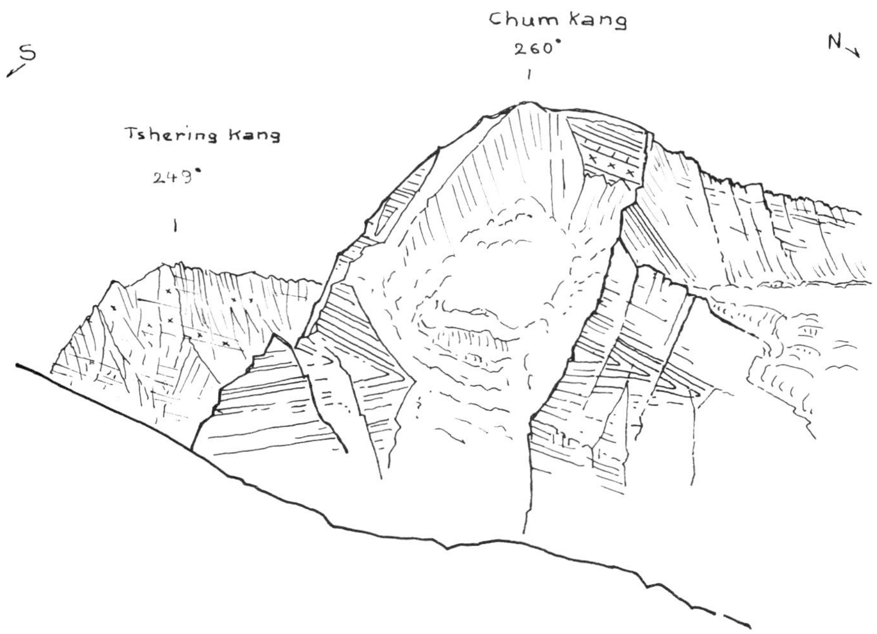

Fig. 117. Sharp folds in marble zones of Chum Kang SW Lingshi. On top intrusions of tourmaline granite. Increase towards SW.

of thickness, which varies from a fraction of 1 m to over 100 m, they may extend for several km. This poses the important problem of their mise en place. Most of the sills do not occur within high grade crystalline rocks, from which they could have been formed by «lateral segregation» with none or only a short travel of the new granitic material. They are emplaced in carbonate rocks which have been altered to marbles or, if impure, to complex calc-silicate marbles. The larger part of the precipitous S walls of the Chomolhari range consist of such a repetition of marbles and granite sills, beautifully displayed in some of the spectacular cliffs (phot. 10). Here again, the contact of the granite sills, regionally concordant to the sedimentary bedding, can in detail be rather wavy, following the slight increases and decreases in the size of the sill. These changes are not reflected by the sediments, which are very sharply cut by the granite. We have already noted how in many places the marbles and calc-silicate rocks expose flowing features, and thin quartz bands are often involved in complicated flow foldings and boudinage (phot. 11 and 12). All flow features are sharply cut by the granites. In spite of this, real crosscutting dykes are rare. They can be seen on the western face of Chomolhari, where a large «folded» dyke of granite cuts through

Fig. 118. The SSW face of Tshering Kang. At the base thick bedded diopside marbles (Chomolhari marble zone) followed by well beddd biotite schists and biotite-psammite gneisses alternating with thin calc-silicate bands and intruded by tourmaline granite sills. The latter increase upwards and northwestwards.

Phot. 69. Isoclinal folds in banded gneisses cut by leucogranite. SE ridge of Chomolhari.

the bedded sediments and gneisses (fig. 23).

The composition of the Chomolhari granites is the same as the main body at Chung La from which they are derived. Aplitic sills are more frequent, however, with less muscovite and more garnet. Here the tourmaline can be locally concentrated in up to 20 cm large radial spots with a white quartz rich rim. Usually the tourmaline crystals are skeletal (phot. 67). The composition, the shape of the granites and their relation to the surrounding carbonate rocks clearly indicate the intrusive character of the granites. We have here an excellent example of the way in which thin (often less than 1 m thick) acid granites have intruded as sills for several km. The surrounding rocks leave no other alternatives. The contact minerals of the limestones, here mostly diopsidic augites and cordierite with a large amount of sphene, give some indication of the temperature involved. No difference can be observed in the composition and shape of the granite sills whether they are intrusive into biotite schists, quartzites or limestones, though the latter rocks are dominant. We can only draw attention to this problem, which invites future detailed investigations.

We have seen how the granites decrease in amount in a northeastern direction. In the more eastern peaks of the Chomolhari range the limestones have considerably increased and the granites become restricted to the higher horizons. This is well visible in the already mentioned Tshering Kang peak which displays a thick foundation of limestones (phot. 70, fig. 118).

Leucrogranites are again widespread in the Lunana area, concentrated in the Gophu La region from where sills and dykes of varied dimensions spread out into the migmatitic gneisses and the carbonate zones. In the upper slopes of the northern border range reaching 7000 m one notes parallel dyke

Phot. 70. The W face of Tshering Kang. Banded marble and calc-silicate layers at the base intruded by leucogranites which increase towards the top with some biotite psammite gneisses.

intrusions similar to the S slopes of the Chomolhari range (fig. 50, phot. 16, 71). Larger granite sills occur however in the southern ranges of Lunana, where thick, rootless sills have already been mentioned. In the glaciated Gophu La region (5400 m) the granites are probably more widespread than what is actually visible. All these granites are here medium to fine grained, rich in tourmaline and often contain larger muscovites with biotite in the centre. Microcline is the dominant K-feldspar and the acid plagioclases are zoned. Garnet is occasionally present. Here again we realize the surprising constant petrology of the leucogranites.

Towards the E and SE the leucogranites intrude the migmatites of the 7550 m high Kankarpünzum, forming a wild net like dyke system rather than the parallel sill type intrusions of the Chomolhari. They also intrude the carbonates of the Zaga La, which we have already recognized as the northwestern extension of the Djüle La metasedimentary belt (phot. 66).

Trough the wild Kankarpünzum range the leucogranite dykes and smaller stocks connect with the Monlakarchung-Pasalum leucogranites which form the largest leucogranite body in the Bhutan Himalaya with an approximate extension of 800 km^2. Its southern and eastern border have already been mentioned while describing the metasediments of the Tsamba area (southern contact) and the Dermalung region (eastern contact). The western contact is exposed in the wildest rockfaces of Bhutan which are the southern walls of the eastern Kankarpünzum and which end in the Melunghi Kang (phot. 41). Here we still recognize the irregular dyke intrusions which can include several smaller stocks. They gradually increase and merge into the main granite body. The peripheral part of these bodies are particularly rich in xenoliths where biotite-psammite gneisses and biotite schists dominate (phot. 72; fig. 122).

The Monlakarchung-Pasalum leucogranite extends northwards into Tibet. After an interruption by a belt of biotite schists, we recognize again dykes and smaller granite bodies in the southern walls of the next and higher range in southern Tibet, the 7600 m high Künla Kangri, a surprising replica of the Kankarpünzum range (fig. 119). Larger granite masses are here apparently not developed, as far as could be observed from a distance in this practically unknown mountain range. It seems to form the northern limit of the leucogranites and of the crystalline sheets as a whole. It extends however as a pronounced spur for over 15 km in a NE direction further into Tibet and borders the low hillocky region of the up-

Fig. 122. Leucogranite intrusion into isoclinally folded biotite-psammite gneiss. Melunghi Chu.

per Kuru Chu where black schists seem to dominate, a fact already observed from the Gwen La N of Sengge Dzong in NE Bhutan. This spectacular north-eastern spur of over 7000 m high and unknown mountains which apparently consist of migmatitic gneisses with leucogranite intrusions is of particular interest (phot. 43).

The petrological composition of the leucogranites has not changed. In some samples the biotite has increased without however surpassing the content of muscovite. Locally the biotite can be chloritized with the formation of sagenitic ilmenite. The tourmaline shows a skeletal development and can be concentrated in perfect star like crystal aggregates (phot. 68). This formation is somewhat different from similar tourmaline concentrations observerd in the Chomolhari and Lhüntse Dzong region and is not related to a more aplitic facies. The spectacular contact metamorphism against the marbles has already been mentioned. Locally Cu ore in the form of bornite with malachitic alteration can be noted in some aplitic border zones. Further E, in the Zakar La region the granite is still the same though tourmaline seems to be somewhat less frequent. Muscovite dominates the biotite which again is locally chloritized. A steep fracture system cuts the otherwise very fresh leucogranite and from such a fracture issues the hot spring of Pasalum to the N of Zakar La.

Apart from these larger northern leucogranite intrusions several local and disconnected occurrences are known further to the S. We have already mentioned the leucogranite sills intruded into the marble section at the southern edge of the Lingshi basin. They are well exposed in the Wang Chu valley (fig. 27) and consist of medium grained tourmaline-muscovite granites. The large muscovites contain small central inclusions of biotite. Tourmaline occurs mostly as skeletal grains in irregular patches. In some places the sills show an incipient schistosity parallel to the bedding of the surrounding migmatite gneisses. The same leucogranites were met further to the W, still along the S border of the Lingshi basin. Their weak tectonization is reflected in an incipient mortar texture of the quartzes.

Leucogranites outcrop as thick sills on the western and southwestern side of the Tang Chu basin. Near Chengana on the W side, a large, fine grained tourmaline-muscovite granite sill is emplaced between biotite gneisses (below) and marble layers. All dip towards the NE (fig. 54). Here too, the granite shows a weak schistosity. Further to the S, in the Tang Chu gorge, garnet bearing muscovite granites form discordant bodies. They intrude into quartzitic schists and calc-silicates as well as biotite gneisses. The granites are rather inhomogeneous with irregular aplite-granitic schlieren. They are rich in muscovite. The plagioclases are free of inclusions and not zonar. Microcline forms irregular grains while the frequent quartzes are large and irregular lobate. The tourmaline has a bluish-green pleochroism. Locally a graphic granite texture is visible.

Further to the E, just before Cändebi, steeply northwards dipping sills of a fine to medium grained tourmaline-muscovite granite are in contact with white marbles which already belong to the carbonates of Cändebi. At the eastern end of the Cändebi carbonate zone begin the Tongsa gneisses with an intricate intrusion of tourmaline bearing muscovite-(biotite) pegmatites. Locally dykes of a

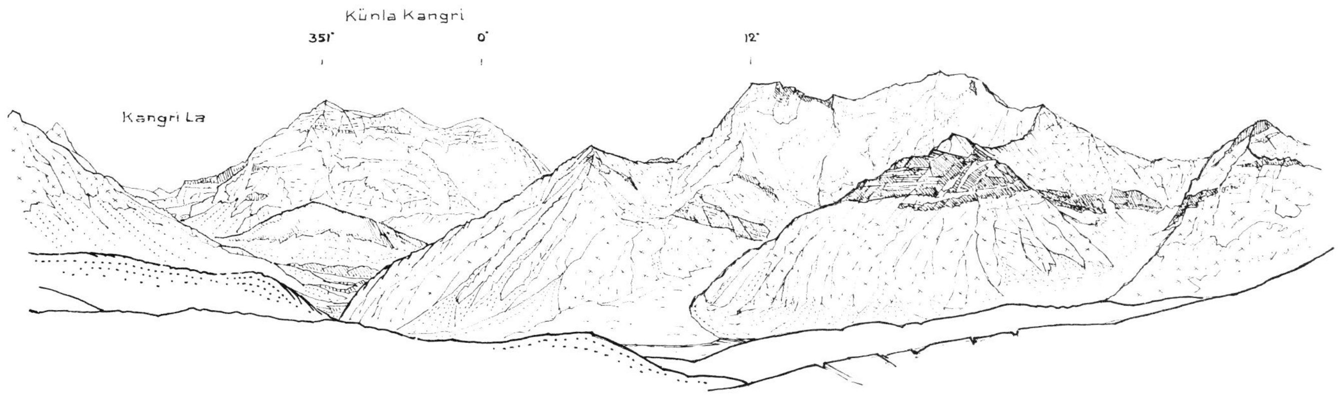

Fig. 119. Künla Kangri (7600 m Tibet) seen from Monlakarchung La. Mountain group to right all tourmaline granite with large xenoliths of dark biotite schists. The same form the base of Künla Kangri which seems to consist mainly of flat lying migmatitic gneisses with younger leucogranite intrusions. Foreground glacier and large moraines with dammed glacial lake. (Very similar to Kankarpünzum).

Phot. 71. The over 7000 m high border range in northern Lunana, with the eastern and western Kangphu-Kang group. In the eastern mountain (right) one recognizes leucogranite sills, alternating with marbles and banded psammite gneisses, similar to Chomolhari. In the western group (left) outcrop migmatitic gneisses with more irregular leucogranites intrusions. They form the base of the eastern group. On top of the latter we note a dark band which could indicate the base of the Tethyan sediments. Peaks in right foreground expose migmatitic gneisses.

Phot. 72. The leucogranite of Monlakarchung La with large xenolith of biotite gneisses and schists. (Peak to the left.) ▶

Phot. 73. The northeasternmost stock of leucogranites in Bhutan, near Bodo La, N of Sengge Dzong. The granite is fractured, similar to the biotite gneisses which it intrudes on its western side. To the east the contact seems against some of the widespread black slates which continue into the W wall of Garula Kang, visible in the right background.

more granitic pegmatite occur without, however, leading to real leucogranite intrusions.

Between the Bumthang region and Tashigang in the E leucogranites are rare. At the top of Rodung La a sill of tourmaline bearing leucogranite is intercalated between biotite amphibolites and platey staurolite-garnet-biotite schists belonging to the metasediments already discussed (fig. 112). S of Lhüntse Dzong, at the head of the northwards ending Kuru Chu -Shumar spur the crystalline zone begins with muscovite-biotite-granite gneisses with irregular intrusions of aplitic garnet bearing tourmaline granites. Here the tourmaline is again concentrated in roundish patches which are rich in larger quartz grains (phot. 67). Northwards the same rock sequence is somewhat sheared in a EW direction and the originally round tourmaline nests are strongly elongated. The contacts of the granite with the granite gneisses is not sharp and differs from the normally sharp and discordant contact of other leucogranites.

In the Tashigang region we noted two local intrusions of leucogranite type rocks. One to the S and near the main thrust, the other within the migmatised section N of the village. Here occur biotite-aplite granites within biotite gneisses which are however not of the typical leucogranite type. Somewhat more characteristic is the granite aplite to the S of Tashigang. Folded biotite gneisses with northeastwards directed fold axis are discordantly cut by a banded tourmaline aplite granite. The frequent microcline has the same rather uniform grain size as the quartz. Tourmaline occurs in layers which may define a primary banding feature. Here again, we seem to have a more special type of the leucogranitic suite.

In the northeastern corner of Bhutan the leucogranites occur only as small dyke like intrusions frequently related to metasediments which have already been mentioned. The dykes consist of fine grained biotite-muscovite granites. Microcline and acid zonar plagioclase is characteristic while tourmaline is rare. Here too, the biotites are partly chloritized. The only exception is a larger stock of leucogranite between the Gwen La and the Bodo La, already mentioned in connection with the metasediments of this remote region (fig. 51, phot. 73). This granite body extends into Tibet and apparently intrudes the black slates, widespread in the upper Kuru Chu. The granite is fractured in a similar way as the gneisses of this wild border range (fig. 51). The NS directed fractures are the latest structural event in this northern area.

With these remarks I end the more descriptive part of the Bhutanese rock sequence and will discuss in the following sections the regional metamorphism and the general tectonic aspect. In this connection I refer to the specially prepared maps (Plate I, fig. 125, 127).

G. The Metamorphism of the Bhutan Himalaya

In the preceding chapter we noted that the Bhutan Himalaya, with its large percentage of crystalline rocks, represents a complicated orogen which exposes old (late Precambrian) crystalline basement with its related metasediments, later biotite granites and subsequently overprinted by a most conspicuous and much younger progressive Himalayan (Alpino-type) metamorphism. Apart from some diaphtoretically affected old crystalline units in the Lower Himalaya, the old rocks of the Higher Himalaya have been fully recrystallized by the neogene Himalayan metamorphic phase. This regional crystallisation is mostly syn- to post-genetic in relation to the main orogeny.

The metamorphic grades discussed in the following represent primarily the results of the latest Himalayan phase and not of the previous Precambrian metamorphism. This is in contrast to the structure which in various places still retains the older trends or reflects a reactivation of an older structural grain.

The metamorphic grades of a complex mountain range can be studied in various ways. In the Bhutan Himalaya three rock associations have been investigated in the field and subsequently studied under the microscope in order to recognize the mineral paragenesis which allows us to deduce the metamorphic grade: 1. the carbonate rocks, 2. the pelitic rocks, and 3. the migmatites and granitized rocks with the youngest melting products, the leucogranites. We will note that the distribution of the latter coincides with the highest metamorphic grades. The three rock types, single and in conjunction, give us the metamorphic picture which is generalized on the included map showing the metamorphic isograds (fig. 125). The results as presented here are still incomplete since they are based on a reconnaissance survey only. However, the main trends show a reasonable picture of the metamorphic grades in the Bhutan Himalaya. Three facts complicate this picture. Firstly the metamorphism is mostly syn- to post-kinematic. Secondly the metamorphic isograds can be either parallel or discordant to the major tectonic trends. Thirdly we note that in many sections the metamorphism is reversed, a fact well known from other parts of the Himalaya. Various explanations for the latter have been proposed, (*Frank et al.*, 1973/1977; *Le Fort*, 1975) which, considering the particular conditions of the Bhutan Himalaya (widespread reversed metamorphism within large crystalline units), do not solve the problem. It certainly remains one of the future main research items.

Apart form the regional metamorphism the young (10–20 my) leucogranites have locally produced a pronounced contact metamorphism.

G.1. The metamorphism of the carbonate rocks

In the Lower Himalaya carbonate rocks occur in the Baxas. Some of the dolomites of the structurally higher horizons in western Bhutan show a sacharoidal structure suggesting a recrystallisation of the dolomite, without, however, any reaction with the siliceous components. In eastern Bhutan, tectonic wedges of Baxa dolomites consist of strikingly isometric dolomite grains, here again suggesting a simple recrystallization. Since the alternating pelitic rocks contain sericite and chlorite in the structurally lower sections and sericite with fine grained biotite in the higher parts, a lower to medium grade greenschist facies must have affected the Baxa carbonates.

Fine grained siliceous dolomite pebbles from the strongly tectonized Diuri boulderbeds in eastern Bhutan, supposedly a stratigraphically younger section than the Baxas, show no reaction at all (phot. 3).

In the Tethyan basins, the stratigraphically highest (Cretaceous) fossiliferous calcareous sediments are nonmetamorphic though they are intercalated in slightly phyllitic black shales suggesting a pre-green schist facies (Lingshi basin). The Paleozoic crinoidal limestones, often disrupted into giant boudins, are still nonmetamorphic at Yale La but grade stratigraphically downwards into fine grained marbles. Towards their base they gradually become coarser grained while diopside and locally tremolite appear. At the basal contact (Wang Chu) widespread calc-silicates are intruded by leucogranite sills. Local contact metamorphism has overprinted the already higher grade regional metamorphism of this lower carbonate section. We have noted already the great similarity with the Chomolhari rocks. Here the high grade decreases towards the E while the carbonate facies increases in the same direction (Dando Gömpa, fig. 30). Flow folding in the basal marble together with the irregular «folded» easternmost aplitic granite dykes still suggest high P/T conditions at the very base of the spectacular limestone wall near Dando Gömpa (fig. 120, 121).

Further eastwards, at Jhari La (fig. 44), at the eastern base of the Lingshi basin, the regional metamorphism has further decreased. Near the contact with the «basement», a coarse muscovite pegmatitic granite, are fine-grained marbles, alternating with some calc-silicates and followed by argillaceous limestones. Upwards, the metamorphic grade decreases rapidly and the overlying shales and limestone are practically non-metamorphic. Very similar conditions were observed at the Toma La carbonates (fig. 49). There the base is complicated by reworked high grade carbonates (with diopside, cordierite, basic plagioclase, biotite) within calc-schists, suggesting an erosion of high grade metamorphic rocks followed by deposits of a medium to low grade metamorphism. Here again the metamorphism decreases rapidly upwards. This general decrease of metamorphism within a normal sedimentary section is characteristic for the whole Himalaya within the Tethyan sediments, and contrasts with the Lower Himalaya and parts of the main crystalline units, where a reversed metamorphism dominates.

In the Tang Chu basin the youngest, still Paleozoic, carbonate rocks are transformed into fine grained marbles of a green schist facies without any formation of new minerals. Only below the Devonian fossil horizon and the conspicuous cross-biotite layers did we note marbles of an amphibolite facies and finally sillimanite gneisses of the basement. It is not certain if the lower marbles belong to intercalations within the basement (Paro type) or form the sedimentary base of the Tang Chu sediments.

All the carbonate rocks within the Paro Series and the related sediments within the crystalline units show a high-medium to high grade metamorphism. The Paro marbles contain a varying amount of diopside and some tremolite. Diopside and garnet are present in most impure marble zones. The carbonate intercalations in NW Bhutan are invariably high grade metamorphic. Beside frequent diopside they contain tremolite, scapolite and often basic plagioclases (bytownite). In the border zone of the large leucogranite of Monlakarchung, contact metamorphism has overprinted the already high grade regional metamorphism with large vesuvianites, wollastonites and frequently garnets. In the Lunana region we observed, intercalated in garnet biotite psammite gneisses, bands of phlogopite-olivine marbles, where the olivine has been partly serpentinized, (fig. 83, phot. 39). A similar incipient retrograde effect is also visible in the surrounding gneisses, where some biotites are chloritized. Retrograde metamorphism is however rare and may have been caused in this area by local tectonization.

G.2. The metamorphism of the pelitic rocks

The pelitic rocks allow a much finer subdivision than the carbonate rocks. They show, except for the Neogene and some Cretaceous and Paleozoic sections, that all the rocks of the Bhutan Himalaya have undergone a more or less intese metamorphism.

In the Lower Himalaya the Permo-Carboniferous Damudas, which are extremely tectonized, show a pre-green schist (anchimetamorphic) facies. A similar weak metamorphism has affected the Diuri boulder slates in southeastern Bhutan, sparing however the carbonate pebbles. The Baxa formation, rich in sericite and chlorite, represents a lower green schist facies. The structurally overlaying Daling-Shumar rocks begin in the Phuntsholing section (fig. 11, phot. 4, 5) with phyllitic slates already containing fine grained biotite, together with sericite and abundant chlorite. We thus note a green schist facies increasing upwards (reversed metamorphism) to high green schist and lower amphibolite facies. Some of the doleritic dykes show the formation of new amphibole. Further up, the amphibolite facies is well documented by muscovite-biotite schists without chlorite, which occur also below the main crystalline thrust.

The same conditions are met in the northern Shumar-Kuru Chu-spur (eastern Bhutan) where biotite-muscovite schists alternate with micaceous quartzites producing pronounced disharmonic folding (fig. 17, 18). Plastic flow folds in some quartzites may also indicate higher PT conditions (fig. 19).

In the Tethyan basin, the pelitic rocks show a gradual increase in metamorphism stratigraphically downwards. This fact, noted already while discussing the carbonate rocks, also contrasts with the sections of the Lower Himalaya.

The Mesozoic and Paleozoic pelites of the Lingshi basin are anchimetamorphic with quartz phyllites grading into a low green schist facies with sericite and some chlorite. Small biotites appear

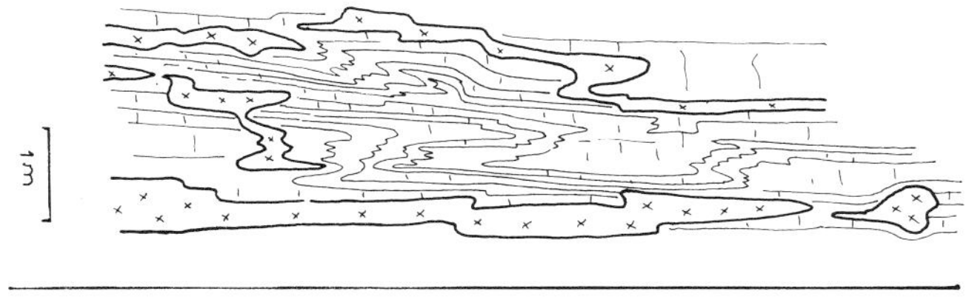

Fig. 120. Flow folded marble with tourmaline granite intrusions. Large marble wall at Dando Gömpa, Lingshi area.

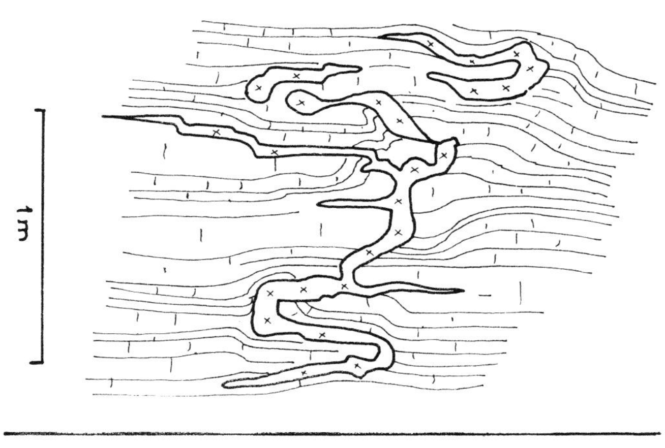

Fig. 121. Detail of «folded» aplite granite bands cutting flow folded marble. Near Dando Gömpa (Lingshi area).

downwards in the chlorite-rich Chekha calc-schists and slates. Towards the base they develop into the wellknown and conspicuous cross-biotite porphyroblastic schists (phot. 8). The porphyroblasts indicate a syngenetic overprint of a low grade amphibolite facies. Below this horizon we enter the thick pile of biotite muscovite schists and gneisses which already contain fibrolitic sillimanite, and which belong to the crystalline units. The change from medium to higher grade metamorphism is gradual and the various horizons are conformable.

In the southern Lingshi basin we noted calcareous beds which increase towards the base, replace the pelitic rocks and are of a high grade metamorphism.

In the Tang Chu basin the sediments reach up into the middle Paleozoic, and include the well known roofing slates of a low green schist facies. The underlying middle Devonian schists, rich in fossils are sericitic. Below a thick section of calc-schists and frequent chlorite schists, still representing a green schist facies follow the well known biotite porphyroblastic schists (phot. 8). They lead to the amphibolite facies with lower biotite-muscovite schists below and higher grade marbles above, intruded by the first sills of leucogranites (fig. 55).

The crystalline units with their intercalated pelitic sediments show a most interesting metamorphic pattern ranging from higher amphibolite to granulite facies. The mineral paragenesis are much more varied and significant than the ones already described from the carbonate rocks.

Every where at the base of the crystalline units is a more or less well outlined thrust and the contact zone exposes rocks from a green schist to lower amphibolite facies below, and a higher amphibolite to sillimanite-granulite facies above. The structures within the crystalline thrust sheets are highly complicated without however a Himalayan penetrative deformation. The tectonic events have been overprinted by a metamorphism which is thus syn- to post-kinematic. Metamorphic facies changes are mostly parallel to the southern thrust contacts but are often discordant to the internal structures. The main mineral association (paragenesis) for the metamorphic pelitic rocks of the crystalline zones are as follows: quartz and plagioclase are always present; K-feldspar is frequent in the migmatites and granitoid rocks; biotite is very frequent and occurs in the following combinations: biotite-muscovite, biotite-garnet, biotite-muscovite-kyanite, biotite-muscovite-staurolite, biotite-staurolite, biotite-muscovite-staurolite-garnet, biotite-muscovite-kyanite-garnet, biotite-muscovite-sillimanite-staurolite-garnet, biotite-muscovite-sillimanite-kyanite-garnet, biotite-sillimanite, biotite-sillimanite-garnet, biotite-muscovite-sillimanite-garnet and locally biotite-(muscovite)-cordierite-andalusite in granites.

The sillimanite-rich zones coincide with increased migmatisation and granitization leading to very irregular granite bodies (often with cordierite such as in the Masang Kang and Domdjen region). The latter must be separated from the younger leucogranites of a deeper seated origin, which also intrude into zones of a high grade metamorphism except in some areas with southern Tethyan sediments (see chemical analysis, and ages p. 132 and 137).

In the crystallines of the Bhutan High Himalaya, as in the entire Himalayan range, the great difficulty remains in distinguishing convincingly the Himalayan from the Pre-Himalayan metamorphism. This is the same problem which faces us when we try to differntiate between the Himalayan and Pre-Himalayan structures. The granitic gneiss bodies in the Shumar-Daling group of the Lower Himalaya (*Jangpangi*, 1974) as well as the Lingtse gneiss from the Darjeeling hills (*Acharyya*, 1975) below the main Darjeeling gneiss sheet are most likely Pre-Himalayan crystalline rocks with only a rather weak Himalayan metamorphic overprint, generally of a diaphtoretic phase. Most of these gneisses derived from porphyritic granites and occur at present as augengneisses with conspicuously large K-feldspar (frequently microclines) crystals. Giant augengneisses of a similar type were also noted in the lower part of the northern crystalline mass such as the gneisses of Tsephu (fig. 111, phot. 58), here however without a retrograde metamorphism.

Here, as elsewhere in the High Himalaya, we regard the main crystalline units as originally Precambrian basement, not unlike the crystalline rocks of the northern Indian shield. This is documented by preliminary Precambrian total rock ages in the Kumaon Himalaya (*Bhanot et al.*, 1977; *Frank et al.*, 1977). The Himalayan metamorphic overprint of the main crystalline bodies culminates in the northern Bhutan Himalaya, where the crystalline thrust sheets have a large extension and where migmatism, granitisation and the ultimate intrusion of leucogranites are concentrated in the higher mountain range.

G.3. The genetic relations of the leucogranites and biotite granites

While discussing the leucogranites in the previous chapters we noted the surprisingly constant petrology all over northern Bhutan and in the other Himalayan regions as well. This fact allows us to recognize this granite type without difficulties in the field and to distinguish it from other Himalayan granites. This uniform petrology is evidently well reflected in the geochemical composition of these granites. The field relations suggest a deeper seated though probably still palingenic source for the leucogranites. They are distinct from the granites formed through granitisation as the final stage of a progressive migmatisation. This fact is visible in the present erosion level, while the source of the leucogranites is concealed in deeper layers. Migmatites and derived granites are clearly cut by the leucogranites. The petrological differences between the older palingenic granites (Masang Kang type) and the leucogranites has been already mentioned. These differences are further substanciated by the following chemical analyses. For these and their interpretations I thank *Volker Dietrich*, Zürich and *Wolfgang Frank*, Vienna.

Geochemical investigation of the leucogranites and biotite granites (Masang Kang type)

Twenty-three granitic samples have been choosen for detailed bulk chemical and isotopic analysis. On the basis of modal and chemical composition two groups (table 1–3) can be distinguished: tourmaline and garnet bearing muscovite leucogranites and biotite granodiorites and granites.

Table 1. Leucogranites

Sample No. weight %	GH 89	GH 93	GH129	GH 170	GH 183	GH 195	GH 197	GH 202	GH 217
SIO_2	74.07	72.19	72.54	73.16	69.27	72.49	73.66	73.43	73.20
TIO_2	0.04	0.13	0.11	0.11	0.23	0.10	0.07	0.09	0.11
AL_2O_3	14.50	15.51	15.37	15.45	16.54	15.64	15.03	14.97	15.07
FE_2O_3	0.53	0.35	0.36	0.32	0.44	0.29	0.14	0.28	0.05
FEO	0.10	0.45	0.50	0.50	1.40	0.40	0.50	0.60	0.85
MNO	0.05	0.01	0.02	0.02	0.04	0.03	0.02	0.02	0.03
MGO	0.04	0.12	0.13	0.09	0.37	0.08	0.07	0.09	0.12
CAO	0.44	0.92	0.77	1.04	0.52	0.69	0.83	1.19	0.76
NA_2O	4.41	3.77	3.64	3.29	3.82	4.70	3.88	3.30	3.57
K_2O	3.83	4.86	4.87	5.06	4.82	3.76	3.76	4.68	4.82
P_2O_5	0.18	0.17	0.17	0.15	0.19	0.16	0.08	0.10	0.16
H_2O+	0.38	0.65	0.84	0.87	1.33	0.85	0.95	0.91	0.98
CO_2	0.00	0.00	0.00	0.00	0.00	0.00	0.00	0.00	0.00
total	98.57	99.13	99.32	100.06	98.97	99.19	98.99	99.66	99.72
ppm:									
BA	45	334	369	208	132	258	128	209	227
RB	237	327	313	336	522	289	268	293	428
SR	8	108	100	69	34	128	34	65	51
PB	11	63	77	77	108	148	53	86	69
NB	<3	<3	3	9	19	3	<3	5	<3
LA	<15	<15	<15	<15	19	<15	<15	<15	<15
CE	<10	20	11	13	26	<10	21	17	<10
ND	<10	<10	<10	<10	<10	<10	<10	<10	<10
Y	<3	<3	15	17	8	4	12	11	4
ZR	<3	54	37	47	43	37	24	39	42
V	<10	11	<10	<10	23	<10	<10	<10	<10
CO	15	13	19	11	14	18	18	16	15
ZN	7	40	38	32	116	43	28	38	46
GA	5	18	15	20	27	13	17	18	20
SC	<1	2	2	3	6	1	2	3	2
modes									
Quartz	32	30	31	32	27	31	35	34	32
K-feldspar	23	27	27	29	26	22	19	24	27
Plagioclase	38	36	34	32	35	43	35	33	33
Muscovite	3	4	3.5	4	6.5	2	7	6	5
Biotite	2	1.5	2	1.5	1	0.5	2	1	1
Opaques (Ilmenite)	0.4	0.8	0.7	0.6	1.1	0.6	0.3	0.6	0.3
Apatite	0.4	0.4	0.4	0.4	0.5	0.4	0.2	0.2	0.4
Tourmaline (T), garnet (G)		T							
Al/(Na + K + Ca/2)	1.12	1.15	1.16	1.17	1.25	1.21	1.26	1.17	1.15

GH 89,93: E Chomolhari
129: Wang Chu gorge, S.
170: Tang Chu, W.
183: Pele La
195: E Bumthang
197: Domdjen

202: Mangde Chu (W Djule La)
217: S Monlakarchung La
221: N Tsamba, contact.
223: Melunghi Chu
378: W Pasalum area
460: Gonto La SW

471: SE Kephu
513: Zakar La
531: Gwen La
507: Chekha S

Leucogranites:

The most distinctive feature of all leucogranites is their relative homogeneous modal composition. In the Or-Qz-Pl triangle (fig. 123 A, B, C, D) the Bhutan leucogranites fall mostly into the field of other High-Himalayan leucogranites, such as form Rongbuk, Makalu, Api, Badrinath and Manaslu.
The entire intergrowth of quartz, microcline and albite indicates an eutectic crystallization, according to *Ghose* and *Singh* (1977) under 4 to 8 kbar pressures and 630–750°C temperature. The locally cataclastic and weak mylonitic features of quartz and the feldspars as well as the planar fabrics indicate that some leucogranites might have suffered a cold deformation after their latest crystallization.

Chemistry:

The leucogranites, although occurring in an area of ~ 15000 km², do not show significant variations (fig. 124 A, B, C, D). They are with few exceptions

Table 2.

Sample No. weight %	GH 221	GH 223	GH378	GH 460	GH 471	GH 513	GH 531	GH 507
SIO_2	72.48	73.04	71.94	72.96	73.16	73.13	72.32	74.52
TIO_2	0.14	0.12	0.11	0.13	0.11	0.13	0.13	0.05
Al_2O_3	14.82	14.95	14.82	14.80	14.52	15.29	14.92	14.79
FE_2O_3	0.07	0.11	0.54	0.16	0.19	0.16	0.21	0.21
FEO	0.90	0.85	0.52	0.82	0.90	0.85	0.80	0.40
MNO	0.02	0.03	0.04	0.03	0.04	0.02	0.03	0.02
MGO	0.17	0.16	0.26	0.16	0.14	0.14	0.14	0.16
CAO	0.94	0.84	0.63	0.86	0.74	0.90	0.74	0.66
NA_2O	3.40	3.61	3.87	3.30	3.04	3.55	3.64	4.28
K_2O	5.17	4.86	4.56	5.18	4.98	5.14	4.74	4.22
P_2O_5	0.12	0.14	0.19	0.18	0.16	0.12	0.15	0.13
H_2O+	0.53	0.73	1.60	0.49	0.75	0.61	0.77	1.27
CO_2	0.00	0.00	0.00	0.00	0.00	0.00	0.00	0.00
total	98.76	99.44	99.08	99.07	98.73	100.04	98.59	100.71
ppm:								
BA	376	274	240	288	243	322	315	140
RB	313	381	416	365	373	326	374	414
SR	85	61	59	76	62	80	68	49
PB	90	72	73	64	45	101	102	59
NB	<3	<3	14	<3	3	<3	<3	3
LA	<15	<15	<15	<15	18	<15	<15	<15
CE	18	18	10	14	34	17	<10	<10
ND	<10	<10	<10	<10	15	<10	<10	<10
Y	3	8	8	6	14	8	4	8
ZR	48	54	55	50	49	53	39	24
V	<10	<10	<10	<10	<10	<10	<10	<10
CO	16	16	22	18	17	19	14	<8
ZN	46	44	62	39	31	48	47	31
GA	15	19	21	18	14	16	16	17
SC	3	1	2	2	3	1	2	1
modes								
Quartz	30	32	30	32	34	31	31	32
K-feldspar	31	28	27	30	28	30	28	25
Plagioclase	34	33	34	31	28	33	34	38
Muscovite	2	3	4	1	6	1	3	4
Biotite	1.5	2.5	2	4	2	5	2	<1
Opaques (Ilmenite)	0.4	0.5	1.0	0.5	0.5	0.5	0.5	0.4
Apatite	0.3	0.3	0.4	0.4	0.4	0.3	0.4	0.3
Tourmaline (T), garnet (G)	T	T		T	T	T	T	T
Al/(Na + K + Ca/2)	1.11	1.13	1.13	1.11	1.11	1.13	1.14	1.13

Table 1,2,3, Bulk rock and modal compositions of leucogranites and biotite granites from Bhutan (analysed by V. Dietrich with x-ray fluorescence methods: *Nisbet et al.* 1978 and *Honegger et al.* 1982).
Cr< 10ppm, Ni < 6ppm, Cu <6ppm, U < 5ppm, Th < 8ppm, S < 20ppm.

(GH 89, 195 and 197) potassium rich granites: for SiO_2 between 71 wt. % - K_2O 4.7 and 5.2 wt. %. A differentiation trend cannot be recognized. This is in extreme contrast with the calc-alkaline granitoids from the Cretaceous and Tertiary Trans-Himalayan batholiths (*Honegger* et al., 1982) which are composed of gabbros, diorites, quartzdiorites, tonalites, granodiorites and biotite granites. Few sensitive and discriminative trace elements, such as Y, Zr, Rb and Ba have been choosen to underline the exceptional chemical difference between the Bhutan leucogranites and the Trans-Himalayan granitoids (fig. 123 A, B, C, D). Although K_2O is only higher by 20 to 30% than in the normal biotite granites, Rb is enriched up to 100% reaching 420 ppm (in GH 183 even 520 ppm). Rb is probably fixed in the micas as well as in the K-feldspars. Muscovite may contain up to 1000 ppm, biotite even 2600 ppm (table 1-3) and K-feldspar approximately 500 ppm (*Hamet* and *Allègre*, 1976). The latter mineral probably constitutes mainly to the bulk rock Rb concentrations. Unusual high, but irregular are Y abundances. This element may be bound in biotite as well as in ilmenite and in tourmaline. All the other trace elements are in general lower in the leucogranites than in the normal calc-alkaline granites. Zr, Ba and Sr show significant lower concentrations by the factor of two. Sr and Ba correlate linearely indicating close geochemical relationships in the feldspars.

Several interesting features emerge from the

Table 3. Biotite-Granites

Sample No. weight %	GH 155	GH 156	GH 158	GH 207	GH 208	GH 336
SIO_2	74.68	71.66	71.15	73.38	73.73	74.00
TIO_2	0.09	0.23	0.25	0.07	0.22	0.25
$Al062O_3$	14.51	14.98	15.34	15.11	13.64	13.81
FE_2O_3	0.00	0.04	0.19	0.11	0.40	0.77
FEO	1.65	1.45	1.55	0.40	1.30	1.12
MNO	0.03	0.04	0.05	0.01	0.03	0.04
MGO	0.10	0.54	0.58	0.04	0.44	0.50
CAO	0.80	1.52	1.69	0.68	0.82	1.03
NA_2O	3.46	3.06	3.38	3.74	4.32	2.86
K_2O	4.66	4.23	3.77	5.58	3.39	5.54
P_1O_5	0.17	0.13	0.11	0.09	0.06	0.10
H_2O+	0.92	0.93	0.84	0.78	0.60	1.20
CO_2	0.00	0.00	0.00	0.00	0.00	0.00
total	101.07	98.81	98.90	99.99	98.95	101.22
ppm:						
BA	150	494	475	363	596	569
RB	375	234	210	291	141	259
SR	58	192	274	83	118	151
PB	76	45	46	131	<6	76
NB	18	<3	<3	<3	4	13
LA	<15	15	18	<15	<15	33
CE	21	37	47	<10	29	42
ND	<10	14	12	<10	14	12
Y	20	6	9	3	12	12
ZR	42	80	99	18	112	40
V	<10	33	39	<10	26	11
CO	8	20	19	15	19	25
ZN	35	49	50	24	10	43
GA	17	14	15	14	7	12
SC	4	4	5	<1	7	4
modes						
Quartz	34	33	32	30	33	33
K-feldspar	25	24	19	30	18	29
Plagioclase	34	35	36	35	40	29
Biotite	2.5	4	6	2	5	6
Muscovite	4	<<1	<<1	1.5	<1	1
Opaques	0.2	0.5	0.7	0.3	1.0	1.5
Apatite	0.4	0.3	0.3	0.2	0.1	0.2
Al/(Na + K + Ca/2)	1.14	1.25	1.25	1.04	1.14	1.04
	S	S	S	I	S	I

GH 155, 156, 158: S Masang Kang
207: Domdjen
208: NE Domdjen (Lamtang)
336: N Sinche La

chondrite-normalized rare earth element (REE) patterns (fig. 124, C). The overall abundances in the leucogranites are lower by the factor of two than in the Trans-Himalayan batholiths (*Honegger et al.*, 1982) The light-to heavy REE fractionation is less evident, except for the granite GH 197, which shows a strong heavy REE depleation. The most remarkable feature is the strong negative Eu-anomaly suggesting that plagioclase has been extracted from the granitic melt. This is consistant with the low CaO-contents between 0.5 and 1 wt.%.

Biotite granites:

Table 3 contains data of six granitic rocks, which belong to the Masang Kang type biotite granites. Biotite is the dominating mica. In addition a minor differentiation trend from granodiorite to granite seems to be present (fig. 123).

Isotope chemistry and ages:

The Rb-Sr and K-Ar data are shown in table 5. The extremely high Rb/Sr ratios as well as the $^{87}Sr/^{86}Sr$ ratios greater 0.75 mark an outstanding contrast to the nomal ratios between 0.704 and 0.706 for granites from the Trans-Himalayan batholith. This is consistant with the results obtained from the Manaslu ($^{87}Sr/^{86}Sr$ > 0.74; *Hamet* and *Allègre*, 1976) and Palung leucogranite ($^{87}Sr/^{86}Sr$ > 0.82; *Allègre* and *Ben Othman*, 1980).

The leucogranites from Bhutan are an extremely

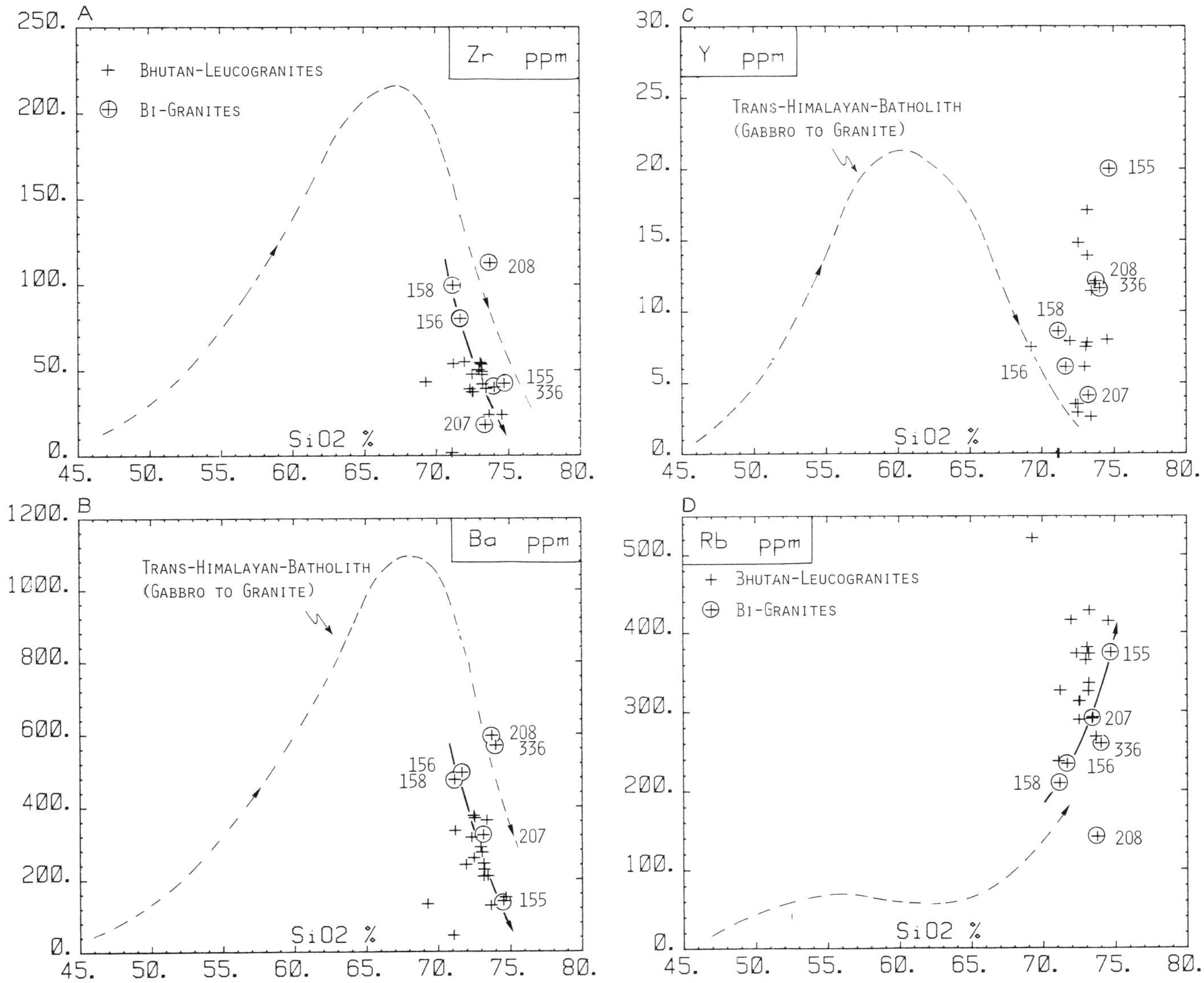

Fig. 123, A,B,C,D. Selected trace element distribution of leucogranites and biotite granites from Bhutan plotted in variation diagrams. The leucogranites show no magmatic differentiation trend. This is in strong contrast to the magmatic evolution from gabbro to granite in the typical calc-alkaline granitoids of the Trans-Himalayan batholith (stippled lines, after *Honegger et al.*, 1982). A possible weak differentiation can only be recognized in the biotite granites.

clear example of granitoids derived from remelted metasedimentary continental crust. Typical calc-alkaline granitoids have not been recognized and related volcanics are completely missing. The rather restrictal acidic leucogranites have intruded migmatitic gneisses and locally contain xenoliths of the host material. The bulk chemistry discussed above indicates typical high potassic granites (K_2O ~ 4.7 - 5.2 wt.%) with low Na_2O contents (~ 3.8 -3.3 wt.%). The Fe_2O/FeO ratios and TiO_2 contents (0.05 to 0.15) are very low and confirm the occurrence of il-menite as well as the low oxygen fugacity of the granitic magma. Plagioclase must have been almost completely removed from the melt prior of crystallization because of low CaO and Eu abundances. Thus the residual rocks must be characterized by the presence of refractory minerals such as plagioclase, hornblende or pyroxene.

The leucogranites are peraluminous in the sense of *Shand* (1950). Using *Chappel* and *White's* (1974) ratio of $Al/(Na + K + Ca/2)$, the leucogranites yield ratios between 1.11 and 1.26. Granites of Trans-Himalayan type in contrast have ratios > 1.1. The reason for the smaller ratios originates from higher Na_2O contents. Only two exceptions with ratios ~ 1.04 have been found within the analysed samples: in the granites form Domdjen (GH 20) and in the biotite granite from Limitang (GH 336), both belonging to the older Masang Kang types.

However, this ratio alone does not favour another source region. Sr-isotopes are necessary to confirm the origin. All the granitic rocks from

A

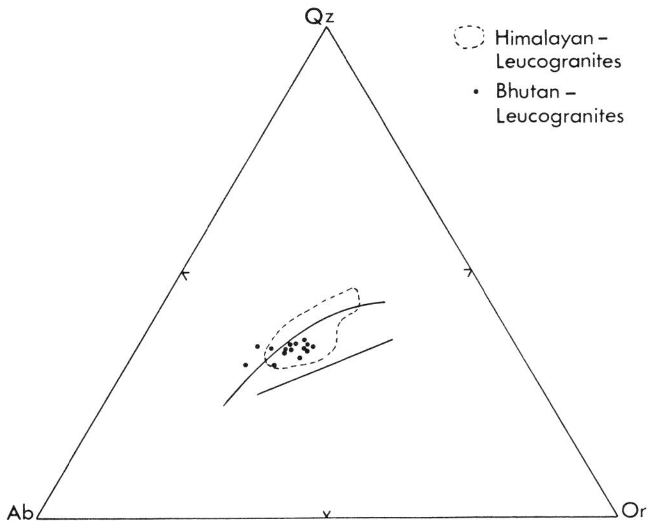

Fig. 124, A. Modal composition of the Bhutan leucogranites within the Quartz-Albite-Orthoclase diagram indicating the close similarity with other Himalayan leucogranites. Line A is the cotectic line along which plagioclase, alkali feldspar, quartz, melt and gas coexist. The numbers correspond to the ternary eutectics at 2, 4, and 10 kbar pressures (*Tuttle* and *Bowen*, 1958). Line B is the cotectic line at higher pressures (*Winkler* and *Ghose*, 1973).

B

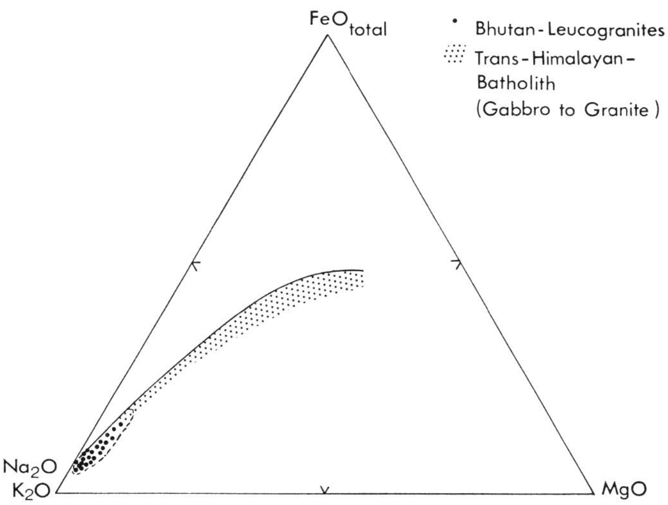

Fig. 124, B. A-F-M diagram illustrating the small chemical variation of the Bhutan-leucogranites compared to the typical calc-alkaline differentiation trend of granitoid rocks from the Trans-Himalayan Batholith.

C

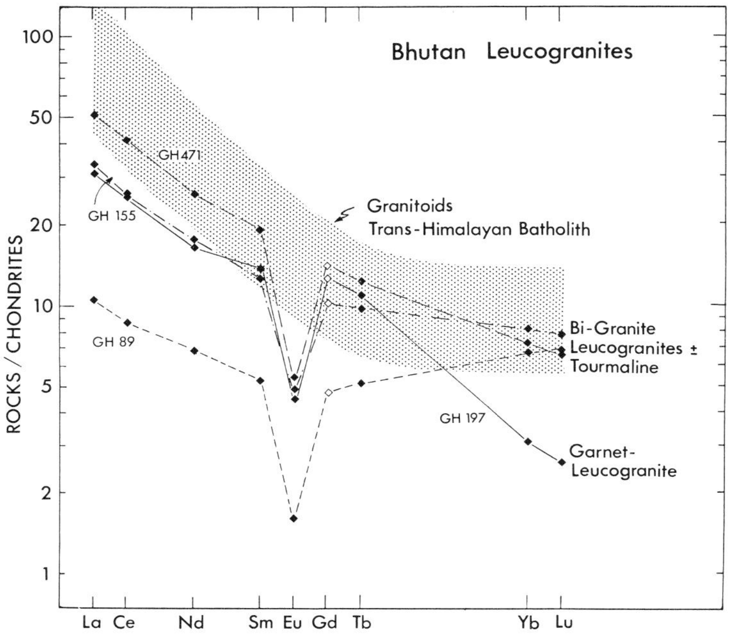

Fig. 124, C. Rare Earth element contents from three selected Bhutan leucogranites and one biotite granite. For comparison field of granitic rocks from the Trans-Himalayan batholith. Values (given in Table 4) normalized to average chondrites of *Nakamura*, 1974.

D

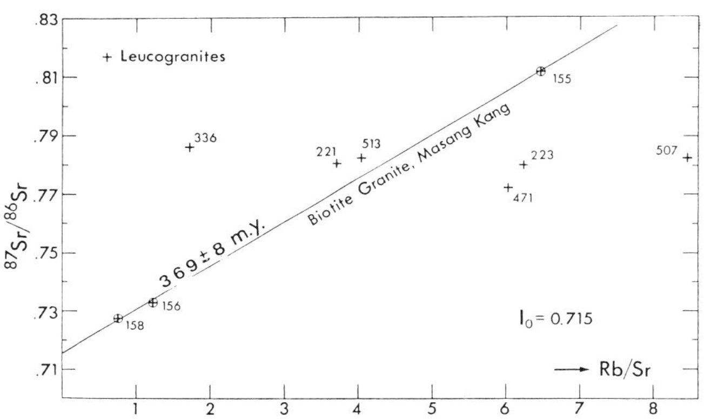

Fig. 124, D. Rb-Sr evolution diagram for biotite granites (whole rocks) from the Masang Kang area. For comparison the late Tertiary leucogranites are shown. They lack isotopic homogenisation.

Bhutan have extremely high $^{87}Sr/^{86}Sr$ ratios greater 0.76. Based on these values together with the other petrographic and geochemical evidence it seems to be obvious to interpret the granites as anatectic melts derived from sialic crust.

Based on field observations, a clear difference does exist within the continental anatectic granites. The biotite granites (Masang Kang type) are developping from highly migmatised crystalline through granitisation with all intermediate stages visible in the field. They are discordantly cut and intruded by the younger leucogranites which originated in a deeper level not exposed by the present erosion surface. While they also derived from a sialic crust their assumed palingenic melt may have formed under somewhat different conditions such as higher temperatures, pressure and fluid circulations related to intracrustal thrusting (Main Central Thrust) (*Ghose* and *Singh*, 1977, *Le Fort*, 1981).

This obvious difference between the leucogranites and the biotite granites is further stressed by the rather unexpected isotopic ages obtained from the biotite granites, mesured by W. *Frank* in Vienna. The total rock Rb-Sr three point isochrone gives an age of 369 ± 8 my, (fig. 124, D and table 5). This age compares to the Paleozoic granites of the western Himalaya with the same Sr-isotope ratio of 0,715 (± 500 my). Since only 3 date point from small specimens of this rock group were available, the calculated age of 369 my may have been reduced, (pers. information W. *Frank*).

These old granites are not related to any well

Table 4.

	GH 89	GH 155	GH197	GH 471
La	3.51	10.6	10.4	16.9
Ce	7.1	22.0	22.3	35.0
Nd	4.34	9.09	10.3	16.6
Sm	1.07	2.58	2.83	3.91
Eu	0.12	0.36	0.34	0.41
Tb	0.25	0.48	0.53	0.59
Yb	1.47	1.80	0.68	1.59
Lu	0.23	0.27	0.09	0.24

Table 4. Rare earth element contents of 3 selected leucogranites and one biotite granite (GH 155) from Bhutan (analysed by A. *Wyttenbach* with instrumental neutron activation technique, described in *Honegger et al.* 1982).

Table 5.

Whole Rock

Sample No	Locality	Rock Type	Rb ppm	Sr ppm	Rb/Sr	$^{87}Sr/^{86}Sr$	Rb-Sr age	K-Ar age
GH 221	Monlakarchung		313	85	3,682	0,78060	–	–
223		Leucogranites	381	61	6,246	0,78004	–	–
471	W Lunana		373	62	6,016	0,77260	–	–
507	S Chekha		414	49	8,449	0,76220	–	–
513	SE Lunana		326	80	4,075	0,78250	–	–
155			375	58	6,465	0,81328		
156	S Masang Kang	Biotite Granites	234	192	1,219	0,73239	369±8 my	–
158			210	274	0,766	0,72762		
336	Limitang		259	151	1,710	0,7859	–	–
305	Thimphu	Migmatitic Gneiss	128	149	0,8590	0,7324		–
		Mineral ages	$^{87}Rb/^{86}Rb$					
507	S Chekha	Biotite	2666	4,20	1904,0	1,0764	11 my	–
		Muscovite	1048	4,93	630,3	0,9455		
336	Limitang	Biotite	1230	1,47	2531,8	1,1541	10 my	–
		Muscovite	579	10,4	167,2	0,8446	11 my	–
305	Thimphu	Biotite	574,9	3,06	549,7	0,8204	11 my	–
542	Tashigang	Biotite	2060	5,57	1093,5	0,9427	13 my	–
		Muscoivte	1095	3,50	938,7	1,0649	26 my	–
186	W Tongsa	Muscoivte (Pegmatite)	–	–	–	–	–	12my
188	W Tongsa	Biotite (Pegmatite)	–	–	–	–	–	11my
330	Paro	Muscovite (Quartzite)	–	–	–	–	–	9±2my

Table 5. Rb-Sr and K-Ar Data

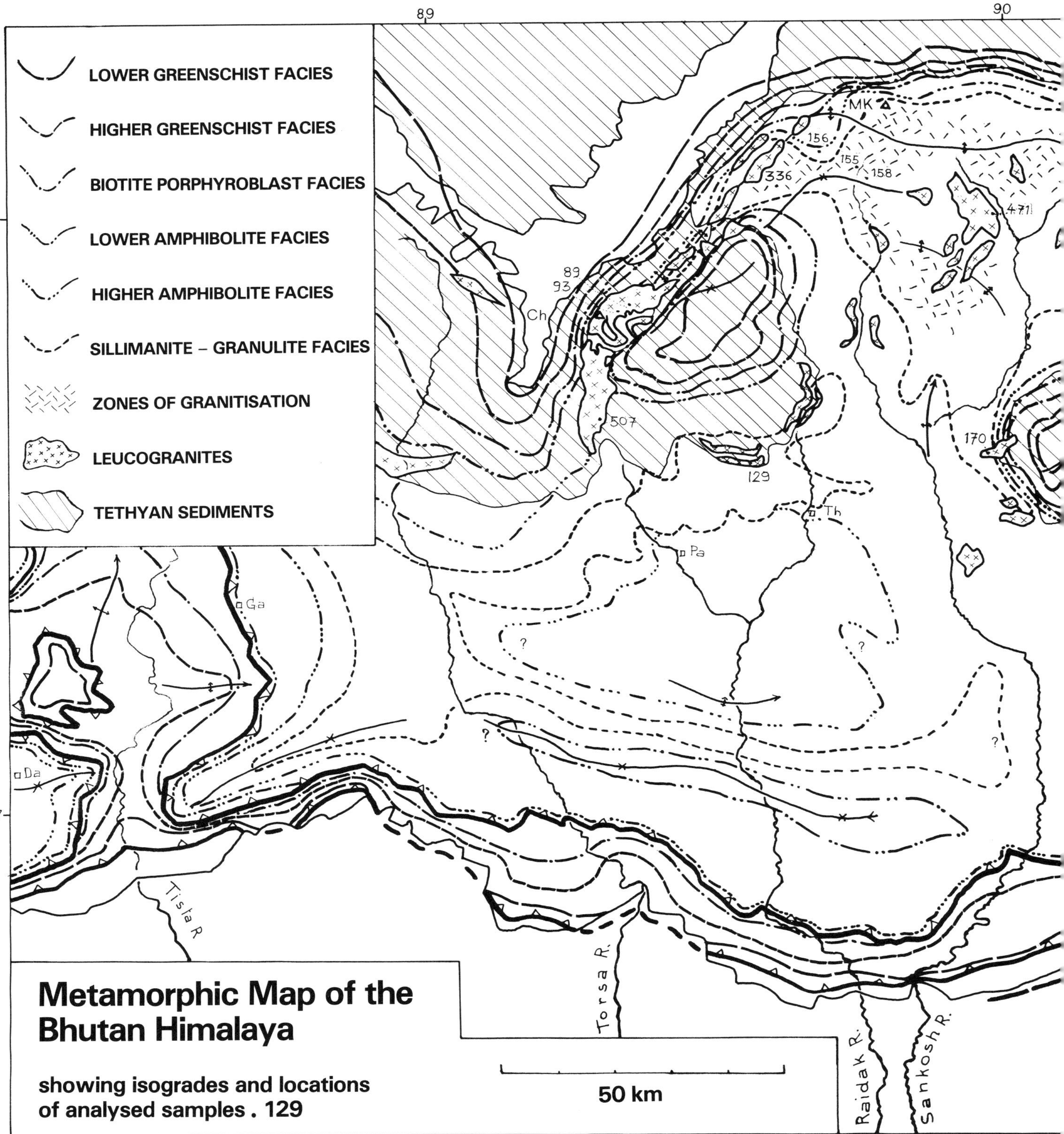
LOWER GREENSCHIST FACIES
HIGHER GREENSCHIST FACIES
BIOTITE PORPHYROBLAST FACIES
LOWER AMPHIBOLITE FACIES
HIGHER AMPHIBOLITE FACIES
SILLIMANITE – GRANULITE FACIES
ZONES OF GRANITISATION
LEUCOGRANITES
TETHYAN SEDIMENTS
Metamorphic Map of the Bhutan Himalaya
showing isogrades and locations of analysed samples . 129
50 km
89
90
28
27
MK
156
155
158
336
471
89
93
Ch
507
129
170
Th
Pa
Ga
Da
Tista R.
Torsa R.
Raidak R.
Sankosh R.

Fig. 125.

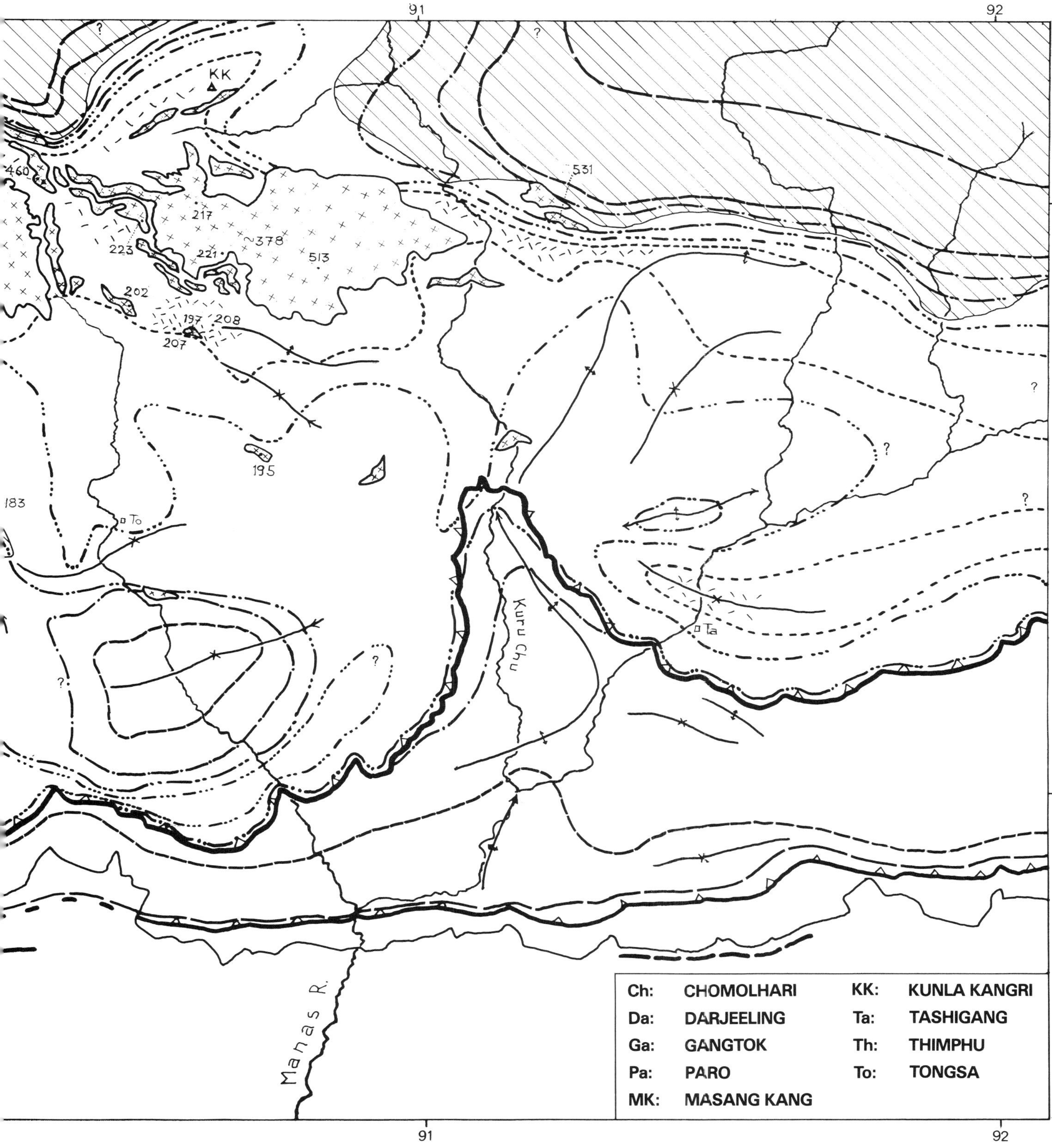
91
92
KK
460
217
223
221
~378
513
202
197
208
207
531
195
183
To
Ta
Kuru Chu
Manas R.
28
27
Ch: CHOMOLHARI
Da: DARJEELING
Ga: GANGTOK
Pa: PARO
MK: MASANG KANG
KK: KUNLA KANGRI
Ta: TASHIGANG
Th: THIMPHU
To: TONGSA

established older Himalayan orogeny but intrude older, mostly Precambrian basement. This interesting fact is a further proof that the crystalline sheets of the Bhutan Himalaya represent Precambrian, locally granitized around 400 my and recrystallized during the much younger Himalayan phase. This phase did not produce any substancial penetrative deformation, and much of the regional crystalline structures represent transported tectonics along the major thrusts such as the MCT.

Contrasting with these older events are the leucogranites with their intrusions subsequent to the highest Himalayan metamorphism. Unusual isotopic features are the high ^{87}Sr ^{86}Sr ratios between 0,762 and 0,782. The Rb-Sr evolution diagram (fig. 124D) shows that the leucogranites do not yield isotopic homogenisation and an isochron for the total rock composition cannot be constructed. The initial $^{87}Sr/^{86}Sr$ ratio seems to be about 0,78, one of the highest figure mesured so far from granites. It can only be obtained by remelting of continental sialic crust. On table 5 we have also shown the mineral ages determinated by *Ruedi Hänny* and *Peter Signer*, the latter's K-Ar ages were already published in Geology of the Himalaya (1964). Muscovite and biotite yield cooling ages of about 11 my, a widespread figure in the Himalaya, ending the syn-to post kinematic main Himalayan metamorphism.

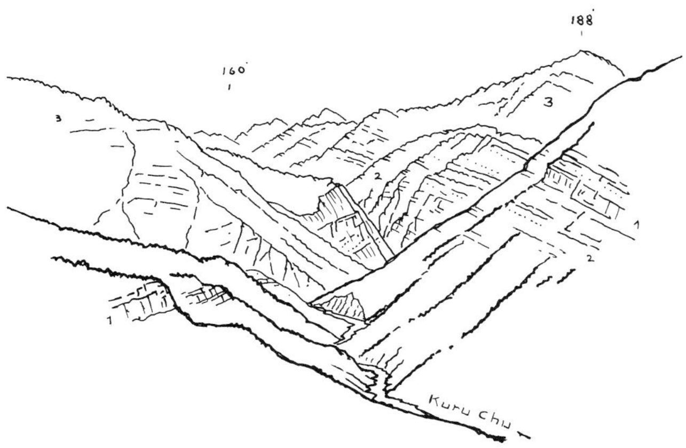

Fig. 126. The northern end of the Kuru Chu-Shumar spur. View Kuru Chu downstream from below Lhüntse Dzong.
1 marbles,
2 quartzites and phyllites below,
3 crystallines sheet.
View towards S.

G.4. Some remarks on the enclosed map with isograds on metamorphism (fig. 125)

No subdivisions are shown within the anchimetamorphic belt, since detailed investigations of the clay minerals were not made. This zone includes the Damudas and part of the Baxa formations as well as the youngest sediments in the Tethyan basins.

The isograds begin with the lower green schist facies as indicated by the appearence of sericite, chlorite and albite. This belt parallels the MBT where it forms the base of a section of reversed metamorphism. In the northern Tethyan basin it terminates the sequence of normally decreasing metamorphism. We draw the isograd of a higher green schist facies with the appearence of fine grained biotite and actinolite in the basic rocks and the growth of small garnets in the schists. A special isograd indicates the first appearence of the conspicuous large cross-biotite porphyroblasts which are a remarkable key bed in most of the Himalayan metapelites (the Budhi schists of the Central Himalaya, *Heim* and *Gansser*, 1939). The amphibolite facies isograd is drawn with the appearence of hornblende, staurolite, kyanite and garnet. Staurolite is a frequent mineral in the metapelites while kyanite is more concentrated in quartz veins and lenses and forms larger crystals. Diopside is frequent in the impure carbonate rocks. A further isograd is drawn with incoming sillimanite, first as fine grained fibrolite but also as larger fibrolitic crystal bundles. It is still associated with staurolite and kyanite and corresponds to a higher amphibolite facies. This belt grades imperceptibly into the sillimanite granulite facies. The isograd is drawn where staurolite and kyanite have disappeared, where sillimanite dominates and where K-feldspar occurs in the strongly migmatized and granitized zones. Garnet is present, locally with cordierite. This goes together with a more or less well developed granulitic texture. Zones with a strong granitization leading to actual granites are shown on the map, together with the younger leucogranite intrusions.

The isograds shown on the map of metamorphism are extrapolated along certain sections. Known spots have been connected with the help of regional structures. Areas where the position is particularly uncertain have been indicated with questionmarks.

H. The Tectonics of the Bhutan Himalaya

The structural information is contained in the preceding descriptive text, the general geological map (Plate I), and is further compiled in the structural map (fig. 127), in the relevant cross sections (fig. 128) and in a generalized profile with a map of the wider structural frame of the Bhutan Himalaya (fig. 136, 137) into the deeper levels. In the following, attention is drawn to items of special interest, beginning in the south.

H.1. The Siwalik belt (Sub-Himalaya)

This belt is bordered by two major thrusts, the Main Frontal Thrust (MFT) in the S and the Main Boundary Thrust (MBT) in the N. The MFT is rarely exposed along the quaternary of the Brahmaputra plain. E of the Tista river and just W of the Bhutan border, the northwards dipping Siwaliks are overriding quaternary terraces which are themselves gently folded and tilted (fig. 4). In south central Bhutan, along the Hatisar (Geylegphu) embayment, sharply tilted terraces coincide with the MFT or a branch of it (fig. 7). In southeastern Bhutan on the S end of the Darranga section (fig. 5), folded Siwaliks are thrust over terraces. In the Kalanodi section further to the E, the contact is obscured by the intervening Diklai boulder beds which may be cut by a branch of the MFT. The exposed fault is certainly of quaternary age (fig. 8).

Embayments of possible tectonic significance we found in the otherwise continuous belt of Siwalik deposits at rare localities along the Bhutanese foothills. The Phuntsholing embayment along the Torsa river is well reflected by a conspicuous bend (orocline) of the Lower Himalaya structures. A similar oroclinal bend is visible in the Hatisar foothills, where I have mentioned the MFT related to tilted and faulted terraces. Two thermal springs are located exactly along this belt (*Jangpangi*, 1976). I have already mentioned the fact that the irregular north front of the Shillong basement (Indian shield) outcrops only 30 km from the assumed MFT and 35-40 km from the MBT. It seems reasonable to assume that some basement spurs influence the foothill belt of Bhutan. The above mentioned Hatisar embayment seems a striking example of such a relation. Along the whole Himalayan front we realize that it is in Bhutan where the shield rocks outcrop nearest to the foothills. A deeper foreland is completely missing here. The MBT to the north of the Siwaliks runs roughly parallel to the sporadically visible MFT and the Siwaliks show no major structural complications, except for some faulting and changes in strike. They mostly dip northwards and rise towards the MFT. An overriding of the Lower Himalaya on partly eroded Siwaliks has already been mentioned along the western Bhutanese foothills (Relief thrust). Post erosional thrusting along the MBT is much more important than so far believed.

H.2. The structures of the Lower Himalaya

Between the MBT in the S and the Main Central Thrust (MCT) in the N is the highly complicated Lower Himalaya. Apart from EW aligned structures we observe large northwards trending spurs. The tectonics of the large Darjeeling spur in the Sikkim Himalaya, adjoining western Bhutan are well known. This picture is somewhat similar to the large Shumar-Kuru Chu spur in eastern Bhutan. First mentioned by *Nautiyal et al.* (1964), the Shumar-Kuru Chu spur shows however more internal structural complications. Both Darjeeling and Shumar-Kuru Chu spurs plunge northwards and may reflect an old but reactivated structural alignement of the Indian shield (fig. 127, 136).

Compared to the western and central Himalaya, the Lower Himalayan belt of Bhutan is narrow and highly compressed. It suffered deformation in response to overthrusting of the thick crystalline sheet of the Higher Himalaya and which extends over the greater part of the Bhutanese territory. The Phuntsholing section shows typical folds and imbrications (fig. 11, 12). The axis of the main lineations, minor and larger folds, plunge against the overriding MCT, and are frequently refolded in subsequent phases paralleling the main thrust (fig. 10). Disharmonic folding played an important role in the thick quartzite sections which contain argillaceous intercalations. The relatively gentle quartzite synforms in the Shumar-Kuru Chu region are a good example (fig. 128).

Phot. 74. The Chomolhari-Masang Kang antiform in the Chomolhari range, seen from the E. Tshering Kang and Chum Kang to the left, Chomolhari to the right. Note the south-respectively north dips in the foreground as well as in Chomolhari main peak.

H.3. The crystalline body of the Higher Himalaya and the Tethyan basins

The crystalline body is over 15 km thick and lies above the MCT. The structural pattern here is controlled by large competent granite-gneiss masses and more incompetent metasediments. The younger Tethyan sedimentary basins are superimposed and the postkinematic intrusions of practically undisturbed leucogranites represent the final phase.

The western crystalline zone is structurally controlled by the Darjeeling spur and the reentrant of the Tethyan sedimentary cover of the Phari region. This depression is flanked east of Phari by the Chomolhari range, which strikes abnormally from SW to NE. Structurally, this outstanding range consists of the Masang Kang antiform which continues northeastwards towards the dominating Masang Kang mountain (phot. 74). From here this structure runs eastwards and forms the high border range towards Tibet (fig. 50). The Chomolhari range is characterised by its profuse intrusions of leucogranites in the form of large plugs, dykes and surprisingly well bedded sills (phot. 10). From the northeastern Lunana region, where the Masang Kang antiform strikes through the tops of the wild border range (phot. 75), the structure swings east-southeast and passes to the S of the largest leucogranite mass of Bhutan, the Pasalum granite. It seems to merge into the northern continuation of the Shumar-Kuru Chu spur. Tethyan sediments transgress along the northern flank of the Masang Kang antiform. They continue into Tibet where they are thrown into east-west striking fold ranges, competent in the Paleozoic and incompetent in the Mesozoic with the incoming flysch facies; (fig. 136, 137).

Apart from this north flank, we have discussed already the Tethyan sediments in the two basins overlying the crystallines, the Lingshi basin in the northwest and the Tang Chu basin in the central part. The Lingshi basin is asymmetrical, with a large, partly imbricated south flank and a short north flank, the

Phot. 75. The S wall of the wild border range in Lunana with traces of the Masang Kang antiform. Irregular intrusions of leucogranites in biotite gneisses with marble bands in the upper part. Migmatitic gneisses in Foreground.

latter cut by the steep Lingshi fault zone (fig. 31, 32, 33). The intense subfolding of the sediments in this basin is mostly EW directed with fold axes plunging gently to the E along its western edge and to the W along the eastwards rising east border. Their strike is sharply cut by the NE-SW directed Lingshi fault which shows a marked upthrow to the NW. The general EW strike of the sediments is still prevalent in the gneiss to the S of the basin. This direction abruptly changes further southwards in the metasediments of the Paro region. Here the fold axes and the major lineaments are NS directed. The dominant NS trend is superimposed on the more generally EW aligned structures on both sides of the Tang Chu basin, and suggests a subsequent phase. This later trend is clearly expressed in the banded gneisses W of Tongsa. We observed it again in the abnormally NW-SE striking banded metasediments of the Djüle La zone where folds paralleling this belt are refolded by structures plunging to the NNE.

We have already drawn attention to the Shumar-Kuru Chu spur with ist complicated core of Lower Himalayan sediments. The northwards plunge of this spur below the crystalline thrust sheet is exposed in the Kuru Chu valley (fig. 126). A northwards continuation of this trend is also reflected in the crystalline which is folded into a complicated and abnormally NNE striking antiform with a conformable NNE plunging secondary fold pattern. This large structure gradually turns eastwards along the Tibetan border in NE Bhutan and takes over the function of the important Masang Kang antiform which we found merging into this spur.

To the E of the Shumar-Kuru Chu spur and its northern continuation reappear structures with a dominant EW strike which continue, gradually changing to the ENE, into the Arunachal Pradesh (NEFA) region to the E of Bhutan.

The structural character, type and direction of folds change sharply along the thrust contact of the MCT. Due to a metamorphic overprint, this contact is not always easy to detect lithologically but can be outlined with the help of structural analysis. The various phases of structural deformation within the crystalline masses are generally older than the main thrusting, except for the reactivated NS directed spurs. Most of the internal deformations suggest transported tectonics of Pre-Himalayan phases.

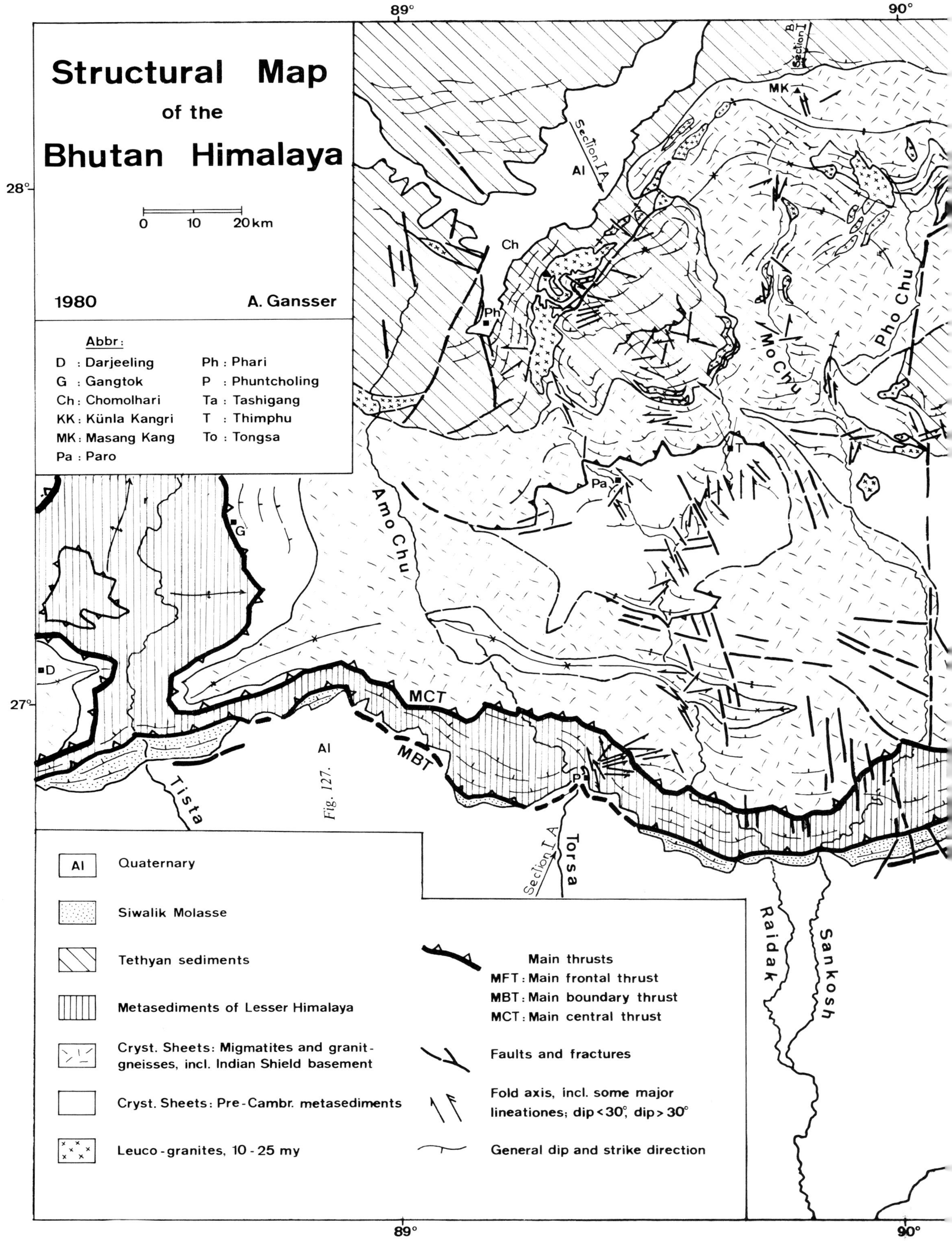

Structural Map
of the
Bhutan Himalaya
0 10 20 km
1980 A. Gansser
Abbr:
D : Darjeeling
G : Gangtok
Ch : Chomolhari
KK : Künla Kangri
MK : Masang Kang
Pa : Paro
Ph : Phari
P : Phuntcholing
Ta : Tashigang
T : Thimphu
To : Tongsa
89°
90°
28°
27°
Al
Section I A
Section I B
MK
Ch
Ph
Mo Chu
Pho Chu
T
Pa
Amo Chu
G
D
MCT
MBT
Al
Tista
P
Torsa
Raidak
Sankosh
Al Quaternary
Siwalik Molasse
Tethyan sediments
Metasediments of Lesser Himalaya
Cryst. Sheets: Migmatites and granit-gneisses, incl. Indian Shield basement
Cryst. Sheets: Pre-Cambr. metasediments
Leuco-granites, 10-25 my
Main thrusts
MFT: Main frontal thrust
MBT: Main boundary thrust
MCT: Main central thrust
Faults and fractures
Fold axis, incl. some major lineationes; dip < 30°, dip > 30°
General dip and strike direction

Fig. 127.

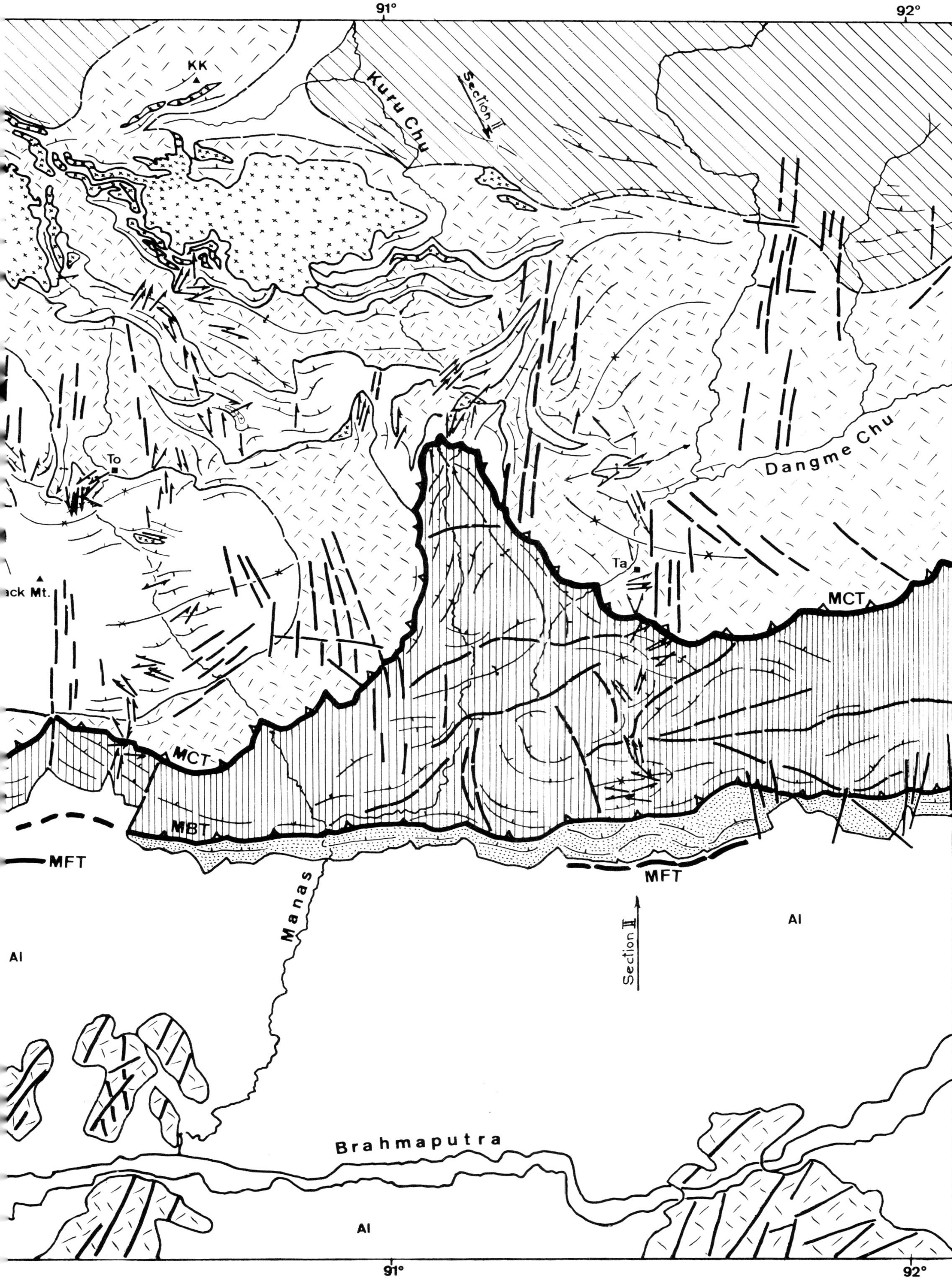

91°
92°
KK
Kuru Chu
Section II
28°
Dangme Chu
To
Ta
ack Mt.
MCT
27
MCT
MBT
MFT
MFT
Manas
Section II
Al
Al
Brahmaputra
Al
91°
92°

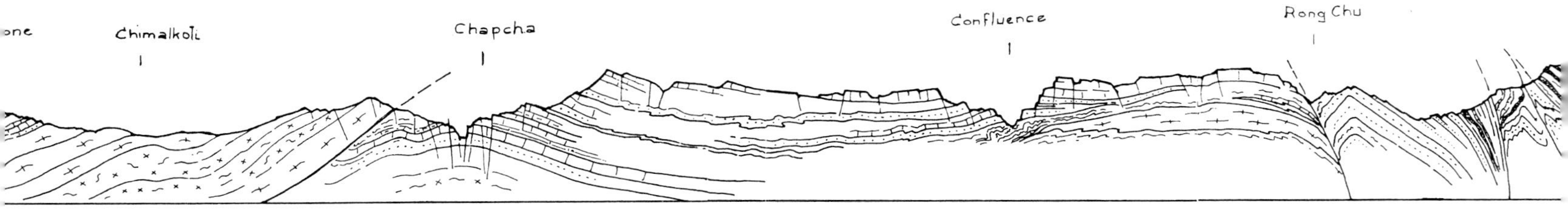
Chimalkoti
Chapcha
Confluence
Rong Chu

Kang Bum
Rodophu Valley
Gasa Dzong

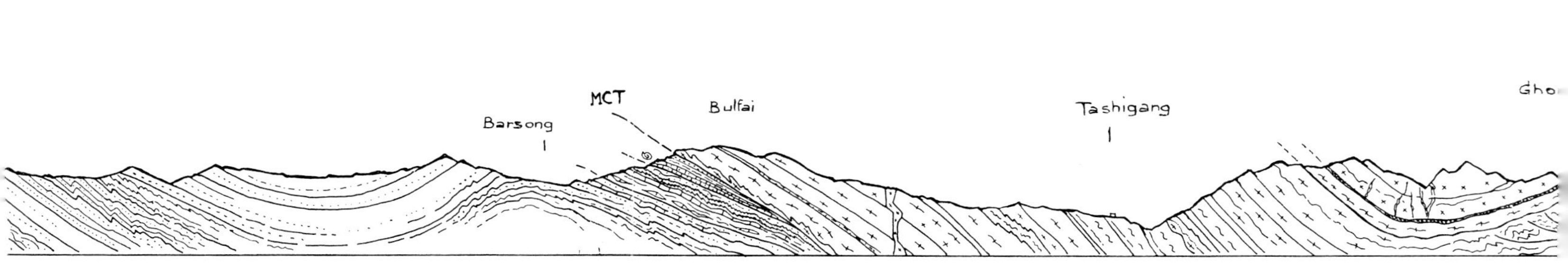
Barsong
MCT
Bulfai
Tashigang

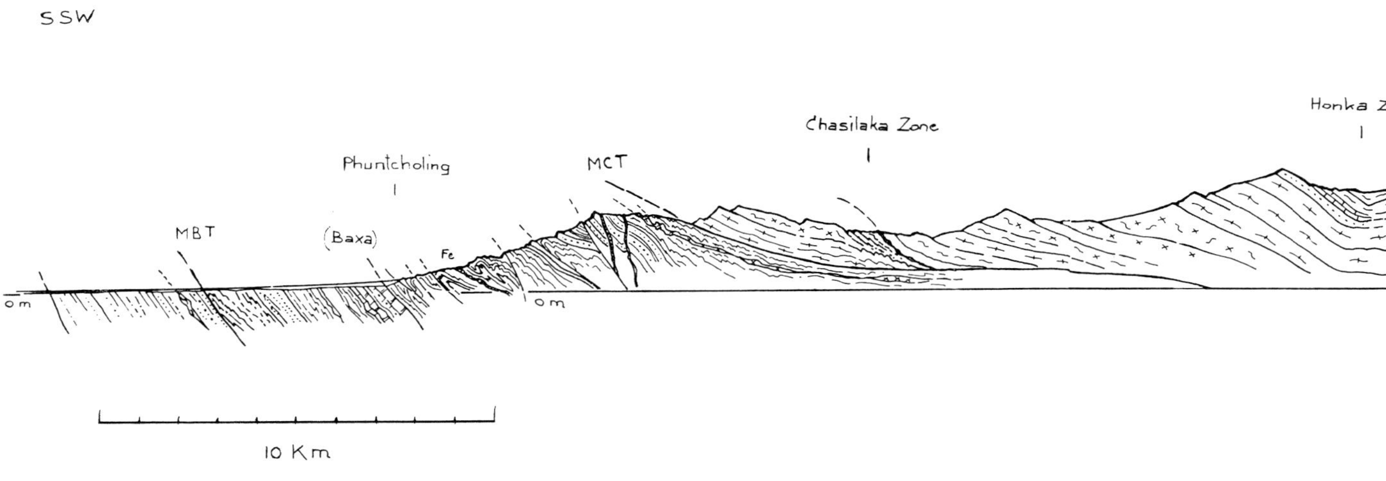

Fig. 128. Geological section I A

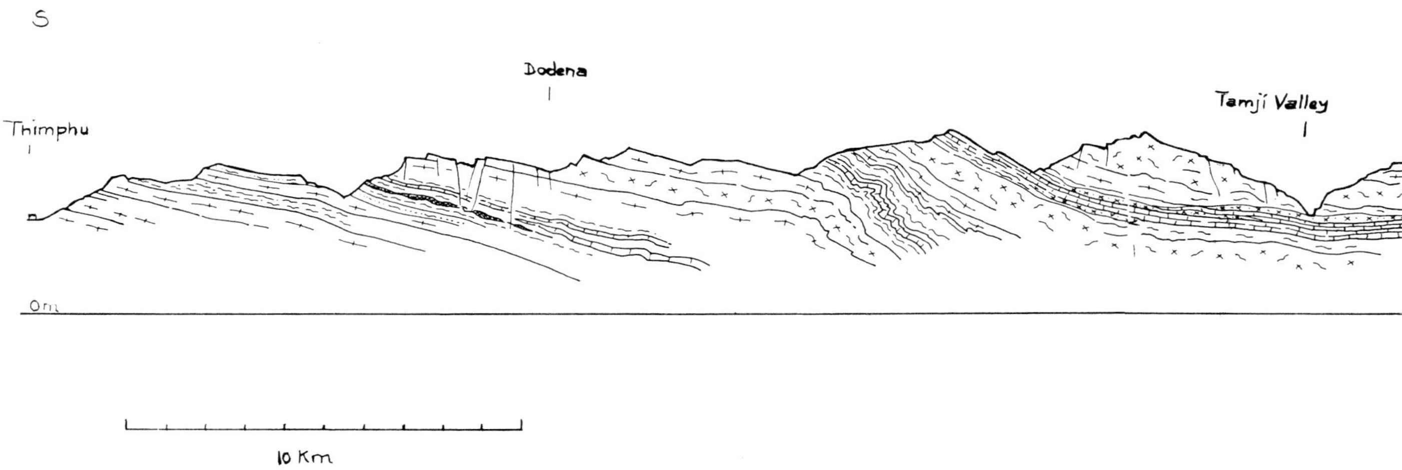

Fig. 128. Geological section I A

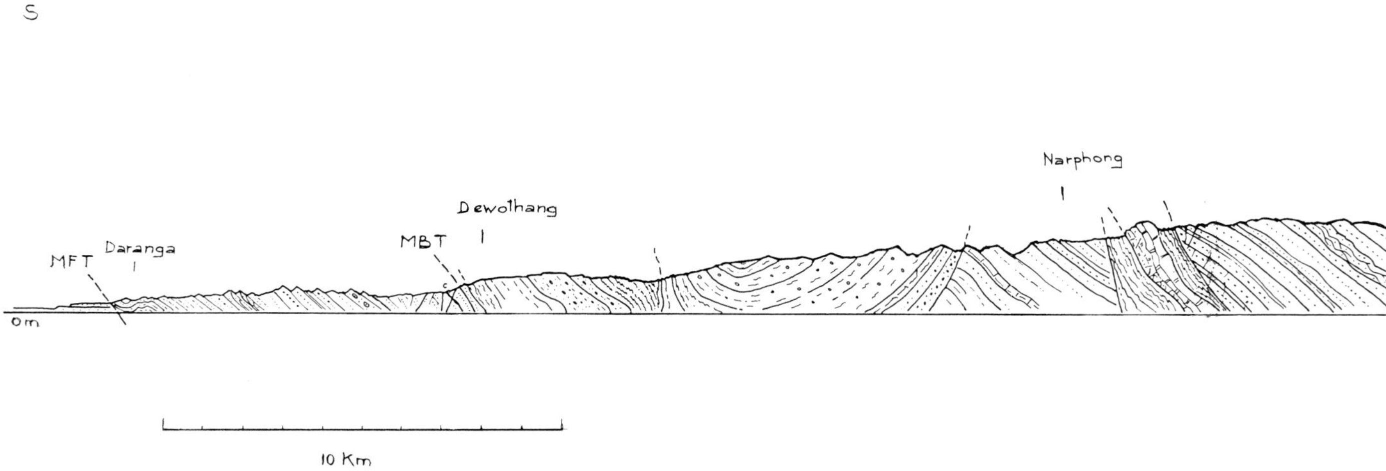

Fig. 128. Geological section II

H.4. Faulting and Fracturing

As a last tectonic phase we note a pronounced faulting and fracturing which has affected and overprinted all previous deformations (fig. 129, 130, 131). It also cuts the Siwaliks and the various major thrusts.

This fault pattern runs predominantly NS with some intermediate NE and SW as well as EW directed alignements. Generally the NS directed faults and fractures are the youngest features and can be detected on the Landsat photos. The faultzone along the Sankosh- Pho Chu rivers and its parallel belt just to the E of the Black Mountains are particularly striking. They may be related to for the abnormally NS directed uplift of the Black Mountains. The fault pattern cutting the northern continuation of the Shumar Kuru Chu spur and the well outlined NS faults within the little known ranges N of Dangme Chu, are also well developed, (fig. 127, Plate I).

In the Bhutan Himalaya, the dominant NS aligned fault pattern seems to represent a reactivated old structural grain of the Indian shield, not unlike the well known Aravalli trend. A similar fault and fracture pattern is visible in the Shillong basement, the northeastern extension of the Indian shield. It is recognizable on the Landsat photos in the crystalline hillocks along the Brahmaputra river, only 35 km to the south of the Bhutanese foothills.

Seismicity seems to be related to this youngest fracture pattern and to its intersection with the youngest EW thrust such as the MFT. Unfortunately the earthquake history and the location of epicenters is still practically unknown within Bhutan. On the new tectonic map of China (1:4000000, 1979) the marked NS directed fault and fracture belt just to the NE of Bhutan is shown to be seismically active with a larger earthquake in 1941. The same area is also striking by a cluster of very hot springs (Liao, 1981).

This young fault and fracture pattern is related to the morphogenic phase which followed the Mio-Pliocene main orogeny. We note a marked change from compressive to tensional features, the latter related to a substancial young uplift (Pleistocene to recent). Based on various sources (*Yang* and *Liu*, 1974; *Hsu*, 1978) one may assume, since earliest Pleistoncene, an uplift of 4000 m with a present uplift rate of probably 5 mm/year, nearly five times that of the Alps. Practically all larger alluvial fans in the main valleys expose a renewed erosion related in most cases to a base level change caused by the recent uplift (fig. 132).

This outstanding uplift is not restricted to the Bhutan Himalaya but may hold for the complete range including the Tibetan highland which amounts to an area of 2500000 km^2. Morphogenic uplifts are well known in all larger orogenies, following the compressive tectonics but are still difficult to explain. The Himalayan/Tibetan uplift is, in its magnitude, unique in the the world.

Phot. 76. Garnet (biotite) fels. Nodular inclusions in Sengge Dzong granit - gneiss to granite. Belongs to calc-silicate type inclusions in Takhtsang-Darjeeling gneisses. Contains almandine garnet with a rim of quartz and plagioclase and some magnetite in the core. Sengge Dzong.

Phot. 77. Garnet-sillimanite-biotite schists. The garnets are idiomorph and free of inclusions. Sillimanite forms bundles, bordered by fibrolite. Domdjen area.

I. Glaciation and Morphology

Glaciation and the consequent strong erosion is strictly related to the remarkable recent uplift of the Bhutan Himalaya (phot. 80, 81). So far no special investigation has been carried out, and the following remarks are based on random observations during the travels in the higher ranges and on the study of the Landsat photos.

Glaciers and corresponding moraines have been indicated on the geological and special glacial map (fig. 133). The outline of the glaciers traced on the Landsat photos and not directly observed in the field is more doubtful, since the available photocover shows a relatively low snow line. For the same reason the mean elevation of the snow line (equilibrium line altitude (ELA), an important date for the regional glacier inventory) could be observed only in a few cases.

Glacial stages: It is most difficult to recognize any early stages below the 3000 m elevation. The intense monsoon on the southern slopes, the present rapide glacial retreat and the often related catastrophic events such as outbreaks of glacial lakes, have destroyed the already badly preserved remnants. This is witnessed by the widespread fluvioglacial sediments, deposited in high terraces and sharply eroded during the present uplift. They are striking by their coarse, partly fanglomeratic composition and reflect the often catastrophic floods. It is a well known fact all over the Himalaya that quaternary terraces and fans are much coarser than the molasse sediments, including even the youngest conglomeratic stages of the Lower Pleistocene Upper Siwaliks.

I.1. Older Glacial Stages

Remnants of the oldest stages were observed in the Koma Chu valley S of Sengge Dzong. At 2900 m elevation a terminal moraine was covered by a landslide. This moraine coincides with a striking change of the valley morphology. Upstream, as well as below this stage the valley is U-shaped and some of the steep cliffs show some glacier polish. In the lower gorge a glacial morphology begins below 2600 m. A similar morphology was observed in the Wang Chu valley N of Cäri Gompa. Here, at 3200 m, the steep valley is blocked by a large spur covered by moraines and further upstream the valley becomes U-shaped. Somewhat more doubtful are scattered moraines on the south side of Sinchu La, E of Thimphu, between 3100 and 3200 m as well as between Thimphu and Pachodin at a similar elevation. A larger landslide complicates the outcrops in this area. Lower stages in the Paro valley at Chuyul between Dukye Dzong and Chekha with terraced moraines at 2700 m elevation and a change into a steeper U shaped valley upstream are also questionable. A similar stage was observed in the Ha valley northeast of Ha Dzong with outwashed moraines at 2700 m.

Correlations with similar stages elsewhere in the Himalaya is difficult. The best known region is undoubtetly the Khumbu valley in E Nepal, on the south side of Chomolungma (Everest) (*Müller* 1980). Even here, stages near the Lukla level, which could correspond to the Wang Chu moraines are highly questionable. Furthermore precipitation in Bhutan is

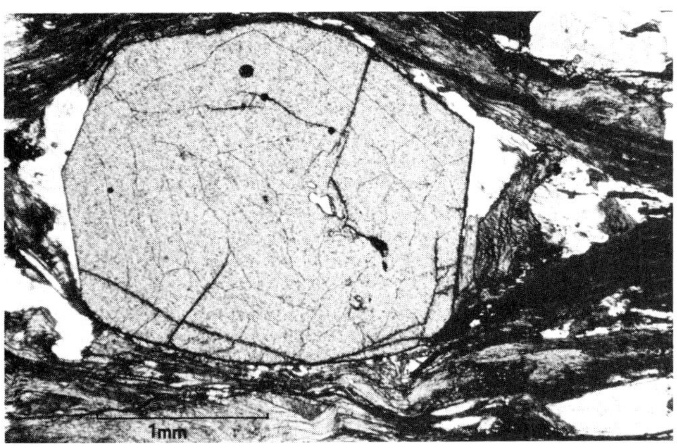

Phot. 78. Idiomorphic garnet, surrounded by fibrolitic sillimanite. Garnet-sillimanite-biotite schists. Domdjen area.

Phot. 79. Formation of fibrolitic sillimanite from biotite. Fine needles growing into quartz-plagioklas grains. Garnet-sillimanite biotite schists. Domdjen area.

Phot. 80. The Domchetang glacier with the Kangphu Kang range. Two lateral moraines exposed: Thanza and recent stages. Central Lunana, view to NE.

higher than in eastern Nepal and the snow line is more depressed. A correlation with alpine stages is again doubtful due to the younger age and larger amplitude of the Himalayan uplift. Also fluctuations during the major glacial periods were more pronounced.

Between these questionable lowest levels and the well outlined moraines belonging to the «Little ice age» the intermediate stages are only partly preserved. Here we can distinguish levels between 3300 and 3400 m, around 3600 m and between 3700 and 3900 m while the «Little ice age» stages vary from 4000 to 4200 m. The latter were dated with the help of lichenometry near the glaciers of the Yatung region (W of Bhutan) with maximum stages in 1818, 1871 and 1885 respectively (*Li* and *Cheng*, 1980).

The lowest intermediate stage is well preserved in the upper Mo Chu valley at Taksaka. A lower stage begins at 3300 m in the valley bottom, followed by well outlined terminal moraines between 3350 and 3400 m at the valley junction where the glaciers from the Laya valley and from the high Masang Kang mountains met. I call the well preserved moraines the *Taksaka stage* while all the lower doubtful occurrences would be *pre Taksaka*.

Moraine levels belonging to the Taksaka stage were also observed in northeastern Bhutan at Dermalung, just to the W of the deep Kuru Chu valley. They show the end stage of a large development of higher moraines. The 3600 m stages have been observed in the Lunana region. In the western valley, between Kephu and Wachey a badly preserved terminal stage occurs at 3500 m. Better developed are moraines E of Wachey, at about 3600 m and belonging to the Domchetang and more eastern glacier, (phot. 80). The lowest visible terminal stage in the eastern valley lies again at 3500 m. This moraine has been strongly eroded by subsequent floods caused by the breaking of glacial lakes. Higher upstream, at about 3700 m, a morainic barrier occurs in the main valley, belonging to a side glacier. All these moraines, between 3500 and 3700 m I would place into a *Wachey stage*. E of Lunana we find the Wachey stage

◀ *Phot. 81.* The deeply entrenched and scree covered Rilo glacier, issuing from the eastern Kangphu Kang mountains. Migmatitic gneisses in foreground with leucogranitic intrusions in the high peaks of the northern border range. Central Lunana, view to N.

again at 3650 m at Tsamba together with old lake sediments and just before the steep gorge of the Chamka Chu. Further to the E, the famous Sengge Dzong basin, filled with lake sediments and moraines at an elevation of 3700 m may also belong to the Wachey stage.

The next stage, between 3700 and 3900 m is widespread and represented by 100 - 200 m high lateral moraines while the terminal ones are mostly washed out by catastrophic floods. Only house sized boulders have been left in place. A large moraine ending at about 3900 m southeast of Lingshi Dzong has been chosen for the type of the *Lingshi stage*. Several large moraines of the Lingshi stage were observed between Sül and Chung La on the southeast side of Chomolhari. They also end at the 3900 m level, again with the typical outwashed end moraines. Between Lingshi and Laya in the NE, the stages between 3800 and 3900 m are particularly well developed. In the valleys between Jhari La and Sinche La the moraines reach 3800 m, sometimes with three very well outlined lateral moraines between 3850 and 4100 m with the uppermost belonging to the «Little ice age» or, the subrecent stage. All these moraines of the Lingshi stage are related to the large glaciers of the Kangcheda group, which dominates the border range between Chomolhari and Masang Kang mountains (phot. 48). In NE Bhutan we find the Lingshi stage well developed in the Pasalum-Dermalung region, where we already mentioned the Taksaka stage. Large terminal moraines end here at 3700 m and are followed upstream by a complicated system of substages, leading to the subrecent moraines. The subrecent and recent moraines seem to merge in the higher parts, above 4200 m, and follow the glaciers up to 5000 m and higher levels. Here one can just distinguish an outer lichen covered moraine and an inner completely fresh band. The subrecent moraines and their frequent substages are particularly well developed in the eastern Lunana region so that this stage between 4000 and 4200 m is called after the highest «village» in Lunana, the *Thanza stage* of 4100 m elevation. Here at least 5 well marked substages are exposed (*Gansser*, 1970). Apart from Lunana, the Thanza stage and substages can be recognized on the south slopes of all larger glacier mountains in northern Bhutan, ending locally in somewhat higher levels, while most smaller glacial cirques have elevations above 4500 m. In the Chomolhari range we note elevations from 4150 m to 4400 m. In the Kangcheda group they are at 4200 m and in the Masang Kang group the stages vary from 4150 m above Taksaka to 4200 m below Toma La with a well separated recent stage at 4700 m. We find large moraines at 4100 and 4200 m in the Rodophu area to the W of Lunana (fig. 85), while E of Lunana and E (and S) of the 7550 m high Kankarpünzum the Thanza stage reaches from 4000 m (Domdjen region) to 4400 m S of the Monlakarchung pass. Excellent glacial stages were observed in the Melunghi Chu valley which runs from Tsamba (Wachey stage) towards the Kankarpünzum peak (fig. 94). The 4050 m level corresponds to the lowest Thanza stage with various higher substages up to 4400 m. Here, the recent moraines begin at 4500 m while the terminal moraine is completely washed out. At the same elevation the large valley glacier ends while lateral glaciers from local cirques are found above 5000 m.

Related to this important Thanza or «Little ice age» stage are the widespread peneplanes between 4100 and 5000 m (fig. 53, 134, 138). They are characterised by «roches moutonnées» interspersed with innuberable lakes. During the «Little ice age» over 100 years ago, these peneplanes must have been presevеd and covered by plateau glaciers, not unlike some small inland ice sheets. We note the same type of peneplane at similar elevations continuing into the Sikkim Himalaya as well as eastwards into Arunachal Pradesh. This type of glacial peneplane seems particularly well developed in the eastern part of the Himalaya. Within Bhutan we recognize four regions where this peneplane has been preserved. In the W between Wang Chu and Paro Chu with an elevation of 4200–4500 m, between Mo Chu and Pho Chu from 4500 to 4800 m and the highest in central Bhutan south of Lunana averaging 5000 m. Further to the E we find this peneplane again between 4200 and 4500 m in the Tsamba area, with its best development in the Zakar La region. The regional picture suggests a large warping of the peneplane level from 4200 m in the western and eastern part to 5000 m in the central part. It is interesting to note that the culmination of the peneplane coincides with the marked NS directed transverse high of Central Bhutan which further to the S is documented by the NS striking range of the Black Mountains, (fig. 135).

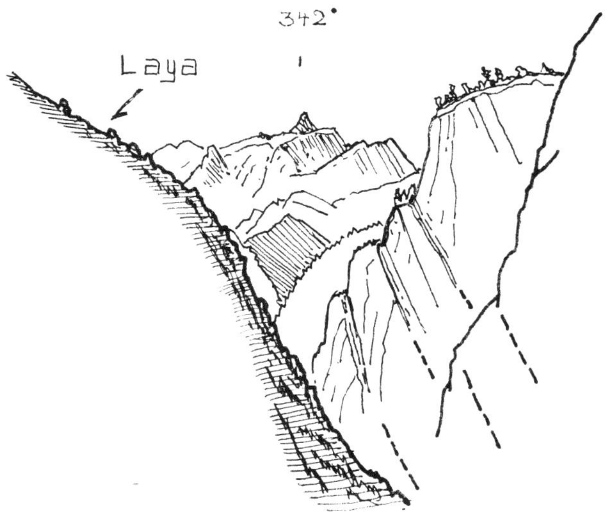

Fig. 129. Steep valley-parallel fault and fracture zones in migmatitic gneisses of upper Mo Chu (above Goyena).

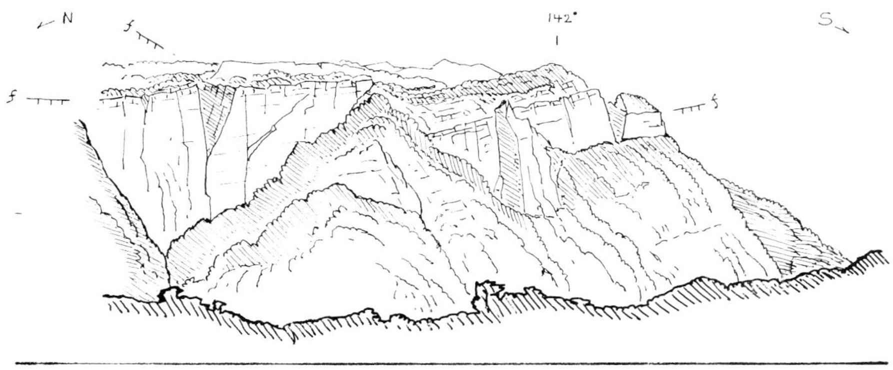

Fig. 130. Fault and fracture zones paralleling the middle Mo Chu opposite Tamji (S of Gaza). The vertical cliffs consist of biotite augengneisses, biotite migmatites with some tourmaline granite mobilisate and in the upper part intercalated calc-silicate and marble bands of Tamji type.

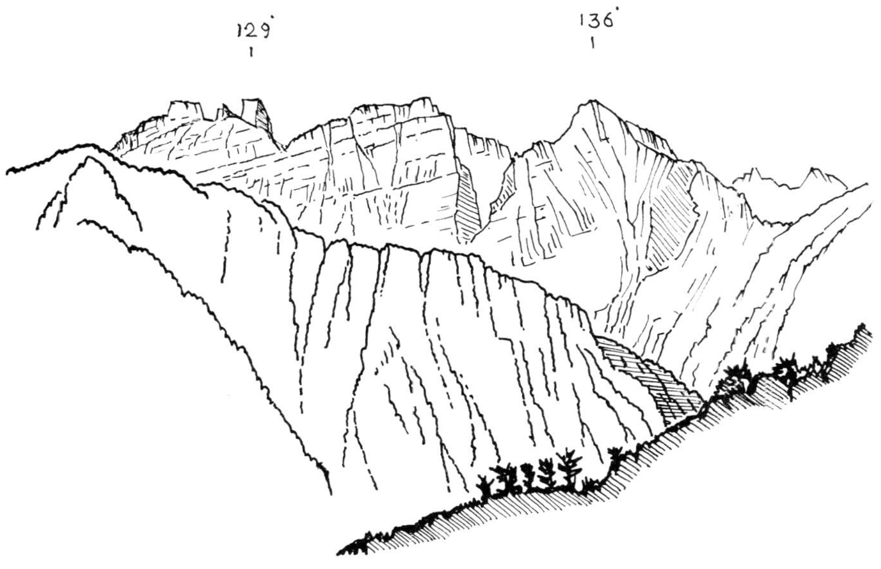

Fig. 131. Regional S rise of gneiss zone to the S of Tsoma La, seen from Goyena (Mo Chu valley). Along valley slope vertical fractures showing in highly jungled rock walls (see also phot. 57).

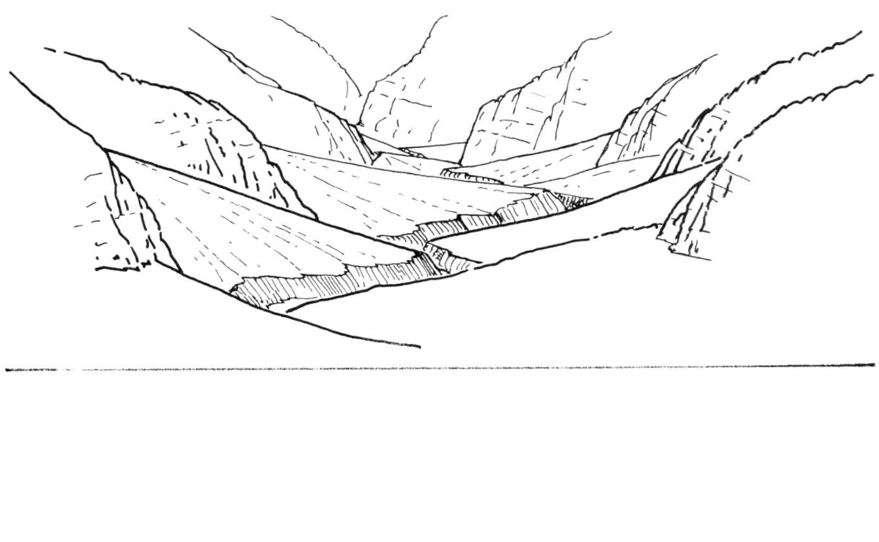

Fig. 132. Renewed erosion of fans in Paro Chu valley between Dukye Dzong and Chugül. View to N. Outcrops on slopes are sillimanite-garnet gneiss of Takhtsang type.

I.2. Recent moraines

They are directly related to the glaciers and begin approximately at the same elevation as the ice level. In larger glaciers a frontal gap of several 100 m up to some kilometers can be noticed, but the heavy morainic load may conceal some dead ice (phot. 81).

On the south side of the highest ranges in northern Bhutan the present glacial tongues reach the following level: Chomolhari 4200 to 4300 m, Kangcheda 4300 to 4400 m, Masang Kang 4500 to 4700 m (fig. 46), Lunana region 4150–4200 m, eastern Kankarpünzum 4400–4500 m, and in NE Bhutan (N of Ora La, Dermalung) 4500 m. All larger glaciers below 5000 m show a strong ablation and the ice surface can be 100–300 m below the lateral moraines (phot. 80, 82). This fact changes above 5000 m which roughly coincides with mean elevation of the firn or snow line which I estimated between 5000 and 5200 m for the high mountains of Bhutan. This could be compared with the ELA values of the better studied southern slopes of Chomolungma (Everest) region (*Müller*, 1959, 1980). In comparison we note in Bhutan a depression of 400 m. The elevations of the glacial tongues is 600 m lower, compared to the larger glaciers of the southern Chomolungma region. The main reason for this depression is largely the higher precipitation rate in the Bhutan Himalaya compared to the Nepal Himalaya. Assuming this condition also for the earlier glacial stages we may compare the stages in the Chomolungma region with the ones established in Bhutan: The Thanza stage (4100 m) could be compared to the «Little ice age» stage of roughly 100 years ago (Dughla stage 4700 m). The Lingshi stage (3900 m) may have its equivalent in the Periche stage (4300 m) while the Wachey stage (3600 m) could compare to the still disputed Pangpoche level of 3900 m. The Taksaka stage (3300 m) seems to have no direct equivalent. It may compare to some of the ill defined levels between Pangpoche and the Lukla stage, the lowest ascertained level in the Khumbu region (*Müller*, 1980). It is interesting to note that the difference in elevation of the corresponding levels diminishes with the increasing age of the stage, assuming the correlation is correct: Thanza/Dughla 600 m, Lingshi/Periche 400 m, Wachey/Pangpoche 300 m, while the present glacial tongues have a difference in elevation between 600 and 700 m. *Müller's* suggestion of a comparatively small scale glaciation during the late or even the entire Pleistocene and Holocene times can be confirmed in Bhutan. This glaciation is far behind the known extensions in the Alps (*Müller*, 1980). It is tempting to correlate these facts with a late but substancial uplift of the Himalaya, still going with an estimate of 5 mm/year.

The dependence of the present glaciation on the precipitation is well documented when comparing the southern slopes with the northern ones. On the south side of Chomolungma (Khumbu region) the glacial tongues of the two largest glaciers, the Ngozumpa and Khumbu, have elevations of 4800 and 4700 m respectively. The largest glaciers on the north side, the Gyabrak and the Rongbuk, end at 5000 and 5200 m. The discrepancy with smaller glaciers is much higher. These facts compare well with the Chomolhari range in western Bhutan, the only region were some reliable data on the northern slopes are available. The largest southern glaciers reach levels between 4200 and 4300 m, while the northern glaciers end at 5300 m. The greater difference is partly due to the fact that no larger glaciers are preserved on the arid northern slopes of Chomolhari.

I.3. Glacial lakes

All the larger glaciers in Bhutan are at present retreating. This is particularly the case in glaciers with tongues reaching below 4500 m. The highly glaciated Lunana region is a good example of this. Some of the lower parts of larger glaciers which are intensely covered by morainic scree deteriorate so rapidly that larger lakes are formed (phot. 83, 84). They are dammed by the recent terminal moraine and, after reaching a certain volume can spill over and break the barrier with the resulting catastrophic flood. Surging of the glacier or avalanching can trigger this outflow. We noted already how in many glaciers the end moraines of the Thanza as well as the recent stages are missing with all the indications of a catastrophic outwash. These events must have happened frequently during the last 100 years, since the strong advance forming the Thanza or «Little ice age» stages. They are happening now and will happen in the near future. (fig. 139). The catastrophic flood of the Pho Chu destroying part of the large Punakha Dzong is a vivid example. I was able to trace the origin of this flood to a glacial lake in the western Lunana valley where an ice avalanche falling from an overhanging glacier into the lake caused the rupture of the terminal moraine which dammed the lake (phot. 85, 86) (*Gansser*, 1970). Ruptured moraines by recent outbreaks of glacial lakes were further observed in the Chomolhari region as well as towards Toma La. Terminal moraines with lakes and which may eventually break, exist in the Lunana region and to the south of Waghye La, (phot. 82). A special report on this dangerous lake situation in the Bhutan Himalaya was handed to the Bhutanese government (*Gansser*, 1978). In the present planning stage of this country it is important to know which valleys may be affected by future flood catastrophies. We have here a striking parallel with the famous «Huaicos» of the Cordillera Blanca in Perù. Only lakes disconnected to larger glaciers can be regarded as safe. They add to the beauty of the landscape, (phot. 87, 88).

Glacial lakes must have been widespread during subrecent glaciations. This is witnessed by the frequent lake deposits, preserved below fluvioglacial terraces and moraines, (phot. 89, 90). Some of the scattered remnants, often high above the present valley floor, indicate larger lakes and suggest catastrophic floods of large dimensions.

Phot. 85. The northern region of western Lunana, N of Kephu dominated by Teri Kang. Catastrophic flood by break of glacial lake (left side of picture), causing large scree of granite and gneisses in valley. This accident caused the partial destruction of Punaka Dzong about 30 years ago (see also phot. 86).

Phot. 86. The near-empty lake after rupture of moraine. (Detail of phot. 85) The glacier above the migmatitic gneiss wall must have advanced rapidly by surging, its front falling into the lake causing the rupture of the moraine damming the lake.

Phot. 84. Large dangerous lake (frozen) formed on lower part of glacier in eastern Lunana. Seen towards northern border range with migmatitic gneisses and a marked fracture system striking NS and dipping towards W.

Phot. 87. A totally frozen, harmless lake near Waghye La with view towards the fractured and migmatitic peaks of the Tsenda Kang range. (E of Masang Kang).

Phot. 88. Some of the harmless larger lakes on Gophu La (5300 m) Eastern Lunana. Note large moraines with iced peaks of leucogranite in background.

87

88

Phot. 89. Lake deposits in the eastern Lunana valley. Large glacier flowing down from northern border range.

Phot. 90. Lake deposits E of Thanza (Lunana), topped by outwashed moraine with boulders, covered by warvy silts.

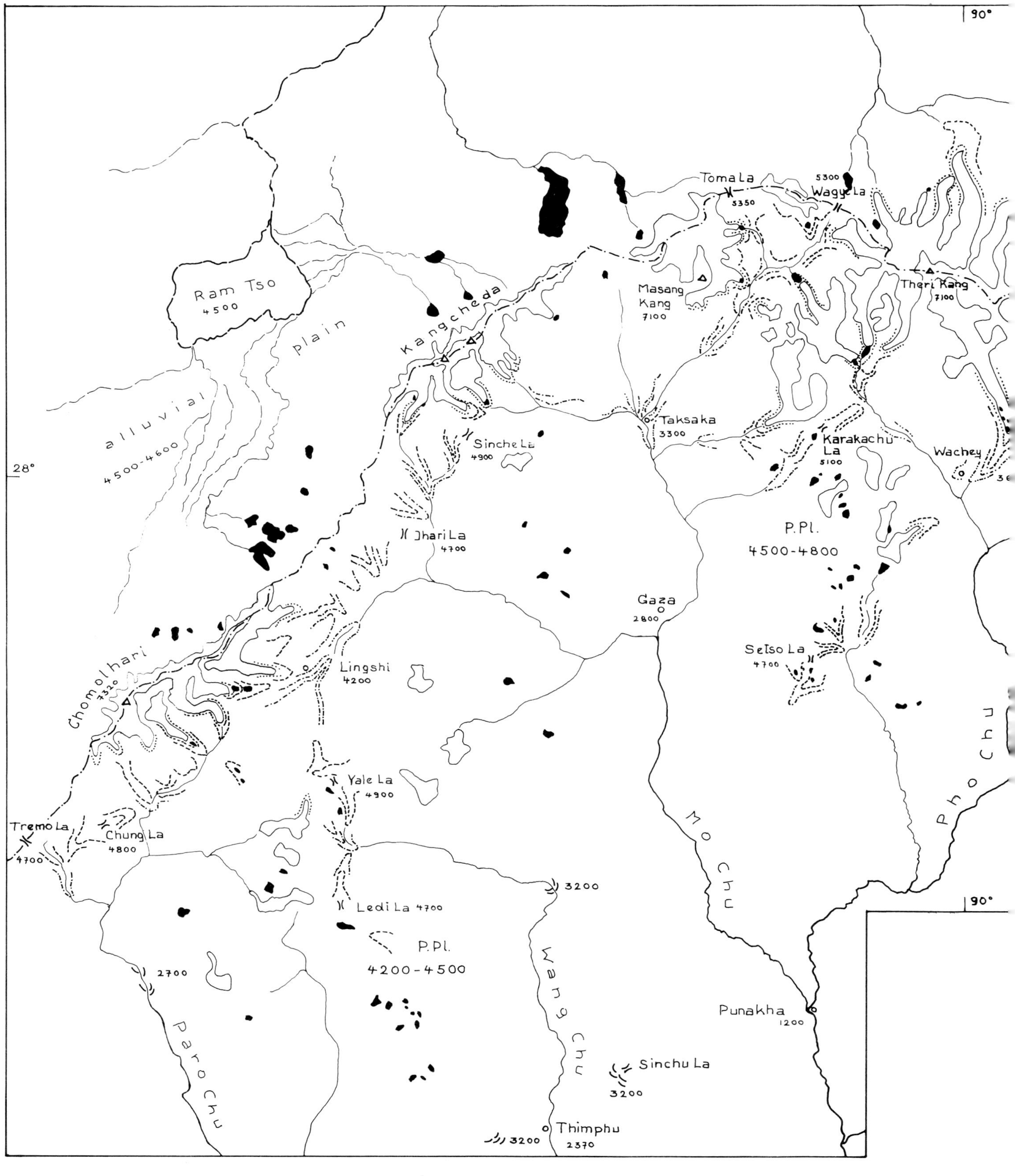

90°
28°
Ram Tso
4500
alluvial
4500-4600
plain
Kangcheda
Toma La
5350
5300
Wagye La
Masang
Kang
7100
Theri Kang
7100
Taksaka
3300
Sinche La
4900
Karakachu
La
5100
Wachey
Jhari La
4700
P. Pl.
4500-4800
Gaza
2800
Chomolhari
7320
Lingshi
4200
Selso La
4700
Yale La
4900
Tremo La
4700
Chung La
4800
Pho Chu
Mo Chu
3200
Ledi La 4700
P. Pl.
4200-4500
2700
Wang Chu
Punakha
1200
Paro Chu
Sinchu La
3200
Thimphu
2370
3200

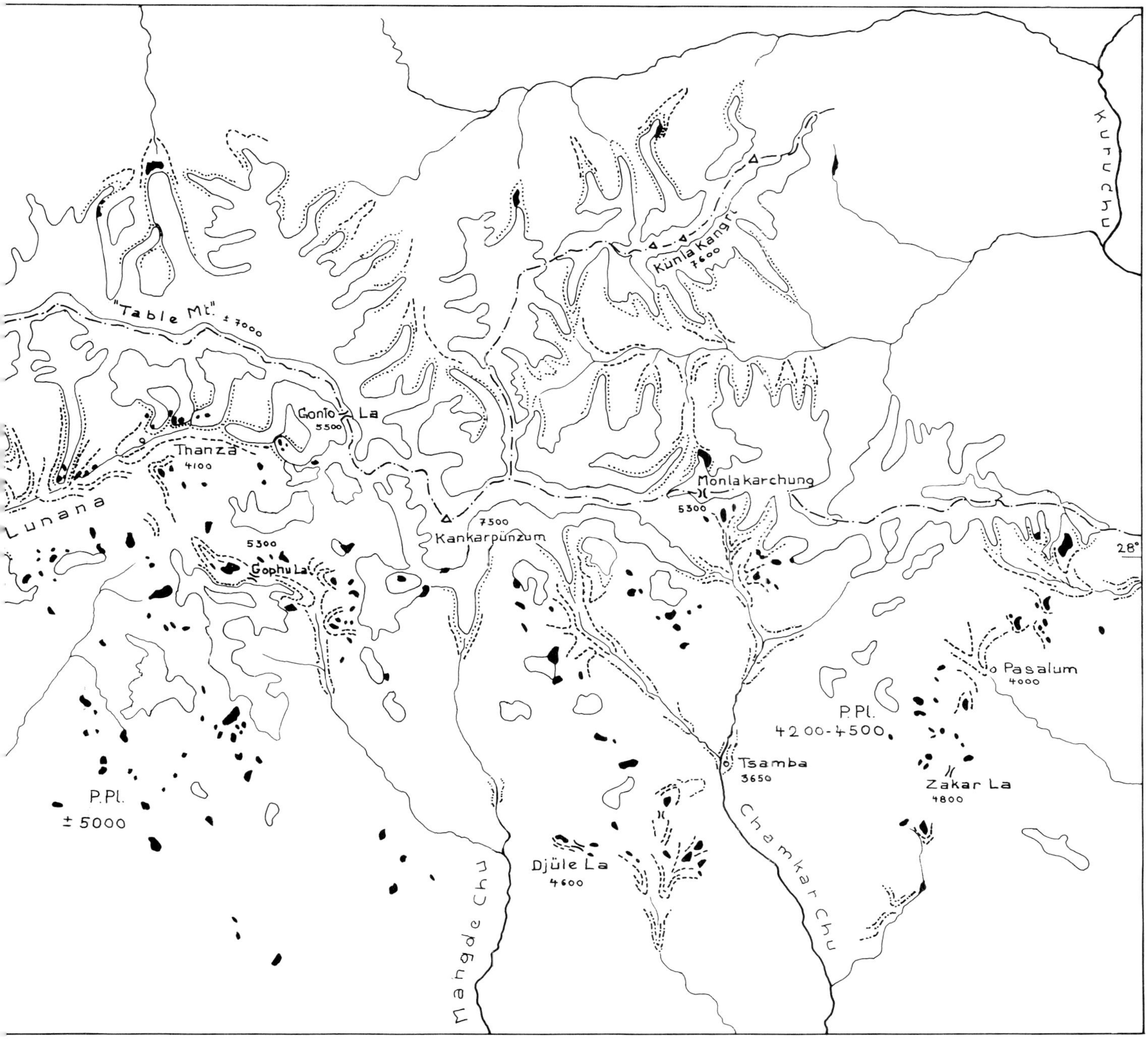

Lakes — Watershed

Present glaciers (End 4200–5000 m)

Recent moraines (4200–4700 m)

Thanza stage (4000–4200 m, ± 150 yrs. old)

Lingshi stage (3800–3900 m)

Wachey stage (3500–3600 m)

Taksaka stage (3300–3400 m)

Older stages (2600–3100 m) p. pl. glacial peneplane

Glacial Map of NW and N Bhutan

20 km

K. The Regional Setting of the Bhutan Himalaya

In the foregoing we have discussed the geology of the Bhutan Himalaya, a slice of about 60000 km^2 situated in the eastern Himalaya. Its structural relation with the Indian shield in the S is well documented, but the westwards continuation into Sikkim is poorly known, except for the foothill zone which has been studied in more detail. Its relations with the Arunachal Pradesh area to the E are not yet fully understood. However, general correlations are possible with the Kameng district, adjoining Bhutan to the E. The northern continuation into Tibet was unknown, but the international Tibet meeting in Peking and the subsequent excursion into southern Tibet had produced many valuable facts (Guidebook, 1980). A map of the wider geological frame-work of the Bhutan Himalaya together with a regional section down to the Moho and including the Suture Zone has been prepared, using all available information. While these contain much data (fig. 136 and 137, some additional remarks are warranted:

In the S the contact of the Bhutan foothill belt with the northern Indian shield rocks is covered by the Assam/Bengal alluvium. We noted however that the northernmost crystalline shield ourcrops are only 30 km from the foothills. This is an outstanding fact unknown elsewhere along the Himalayan chain and shows that practically no foreland basin exists between the shield and the Himalayan orogen. The first, southernmost and youngest structural break is indicated by the sporadically outcropping Main Frontal Thrust (MFT) displacing the Quaternary. It may be related to the regional morphogenic uplift which also caused the NS directed faulting and fracturing of the Bhutan Himalaya and the large N directed embayments such as the Darjeeling and the Kuru Chu-Shumar spur. This predominantly NS aligned trend may reflect a reactivation of the old northwards directed structural grain of the northern Indian shield such as the Aravalli trend which originally has a pre-Vindhyan age (Precambrian). We have already drawn attention to the interesting fact that the northernmost crystalline spur of the Indian Shield, W of the Manas river, coincides with a pronounced gap in the foothill belt in the Hatisar region where the Siwaliks are missing and conspicuous Quaternary terraces are displaced along the MFT.

In the W the Darjeeling halfwindow or spur influences the western part of Bhutan. The structural complications within the main crystalline sheet, such as the western continuation of the Paro thrust, are not yet well known.

Of special interest is the abnormal Chomolhari trend which borders the Chumbi valley depression (Phari Dzong) on its eastern side. NS faults and a very marked N and NE striking monoclinal flank of sediments E of Phari Dzong are striking features. The Chomolhari trend is also outlined by an antiform between the Phari Dzong basin and the Lingshi basin. This antiform swings gradually into a WE direction (Masang Kang region) and limits the main crystalline sheet towards the N. Its northflank is transgressed by the Tethyan sediments continuing into Tibet.

To the E, in the Kameng region, the Siwalik foothills and the Lower Himalayan metasediments are well developed. The latter, beyond the pronounced Shumar-Kuru Chu spur increase in width with various secondary thrusts. This widening may coincide with a larger crystalline uplift in the Tawang region (fig. 138). The MCT is well outlined S of the Sela pass, striking northeastwards. The structural complications further to the E in Arunachal Pradesh region are visible to the E of the Baraili river on the Landsat photos. The young, still active fracture belt

Fig. 134. View from Rodung La to the gently folded crystalline nappes in the Chamka Chu region. *KP* = Kankarpünzum, *CH* = Chamka Chu valley, *KK* = Künla Kangri, *KE* = high range NE of KK.

N of Tawang, in the Kameng/Tibet border region, is of particular interest. These marked NS directed structures seem less frequent further to the E. They parallel the faults in the northern continuation of the Kuru Chu spur which cut NS directed anti- and synforms in the crystalline sheet. Both turn gradually eastwards, the antiform taking over the function of the Chomolhari-Masang Kang structure.

To the N we find the wide, complicated «basin» of Tethyan sediments, between the northern Bhutan antiform of the crystalline thrust sheet and the conspicuous Suture Zone along the Tsangpo river. The latter is bordered to the N by the granitoids of the Transhimalaya (Kangdese belt of the Chinese authors).

The northern border of Bhutan is formed by the Chomolhari-Masang Kang antiform and its equivalent further to the E. It coincides with the highest metamorphism and a Pre-Himalayan well developed partial melting which produced the 400 my old cordierite-bearing biotite granites. These have to be distinguished from the profuse instrusions of the much younger leucogranites which are concentrated in this northern belt. Their origin is still somewhat doubtful, but they certainly derived as palingenic granites from a deep continental crystalline source, which distinguishes them sharply from the granitoids of the Transhimalaya with their more oceanic origin and marked differentiation.

The discovery of the abnormal trend of the spectacular, crystalline Künla Kangri range (the highest mountain in the eastern Himalaya apart from Namche Barwa), which strikes northeastwards and plunges into the Tethyan basin, was of special interest. It breaks the trend of the Masang Kang antiform which seems to pass further to the S, just S of the largest leucogranite intrusion of the Pasalum region.

In the Tethyan «basin» the main structural trends were outlined with the help of the Landsat photos and the Guidebook of the Tibet meeting (1980) gave additional information. The platform sediments, beginning with the Middle Paleozoicum change to the N into a flysch facies, beginning with the Triassic, and this change coincides roughly with a conspicuous range of crystalline outcrops, often associated with young granites. They outcrop to the W of the Yamdrok Tso while to the E we note a marked thrust line, branching off from the Suture Zone. This latter zone is well outlined between Gyantse and Shigatse. In spite of locally excellent exposures, the internal structure of this Suture Zone is still little understood (re. also *Gansser*, 1980).

The southern front of the ophiolitic belt of the Suture Zone is thrust to the S, locally forming nappes. The base consists mostly of an ophiolithic melange belt with large exotic blocks of Permian and Triassic platform limestones but also includes some blocks with a rich Globotruncana fauna, defining the youngest age of the exotics. Ultramafic rocks form, as elsewhere along the Suture Zone, the structurally highest units.

To the N the ophiolites are thrust steeply onto an enigmatic molasse-flysch section, the Shigatse formation of the Chinese authors. It transgresses the Transhimalayan granitoids but contains intercalated pelagic sediments with Upper Cretaceous ammonites, discovered during the 1980 S Tibet excursion. The age of the Shigatse formation therefore corresponds to the age of the Kangdese plutons (Transhimalayan pluton), which varies from 110 (basic) to 40 (acid) my (Guidebook, 1980 and *Honegger et al.*, 1982). The enigma of the rapid uplift and erosion of the plutons and the practically simultaneous sedimentation of the Shigatse formation is still unsolved. The Kangdese plutons intrude the Mesozoic platform sediments to the N of the Suture Zone and seem to form an accretional belt along the «Asiatic» plate.

Great difficulties are met when trying to draw a section through this wider frame of the Bhutan Himalaya down to the Moho level (fig. 137). Three main units can be distinguished by their internal structures and the depth to the Moho. 1. The Indian shield, a typical continental plate, strongly fractured in the upper part and granitized in the lower levels with the Moho between 25 and 30 km. 2. The complicated Himalayan sector, separated at present from

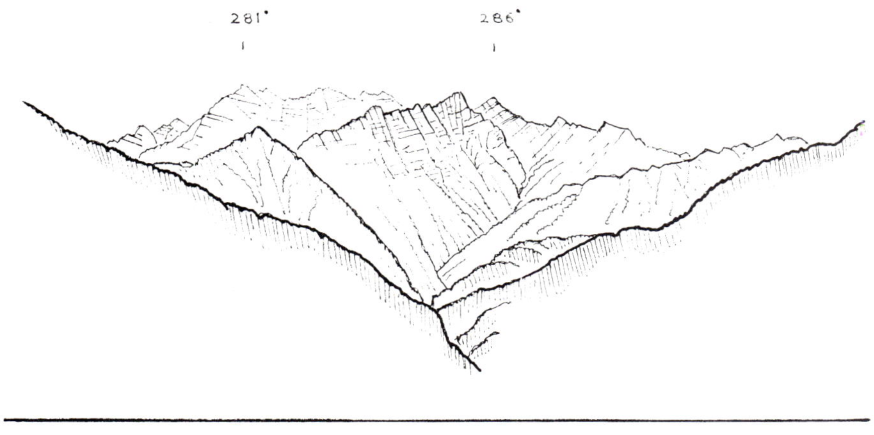

Fig. 135. The basin like tectonics of the Black Mountains seen from N of Shamgong Dzong, (Mangde Chu valley).

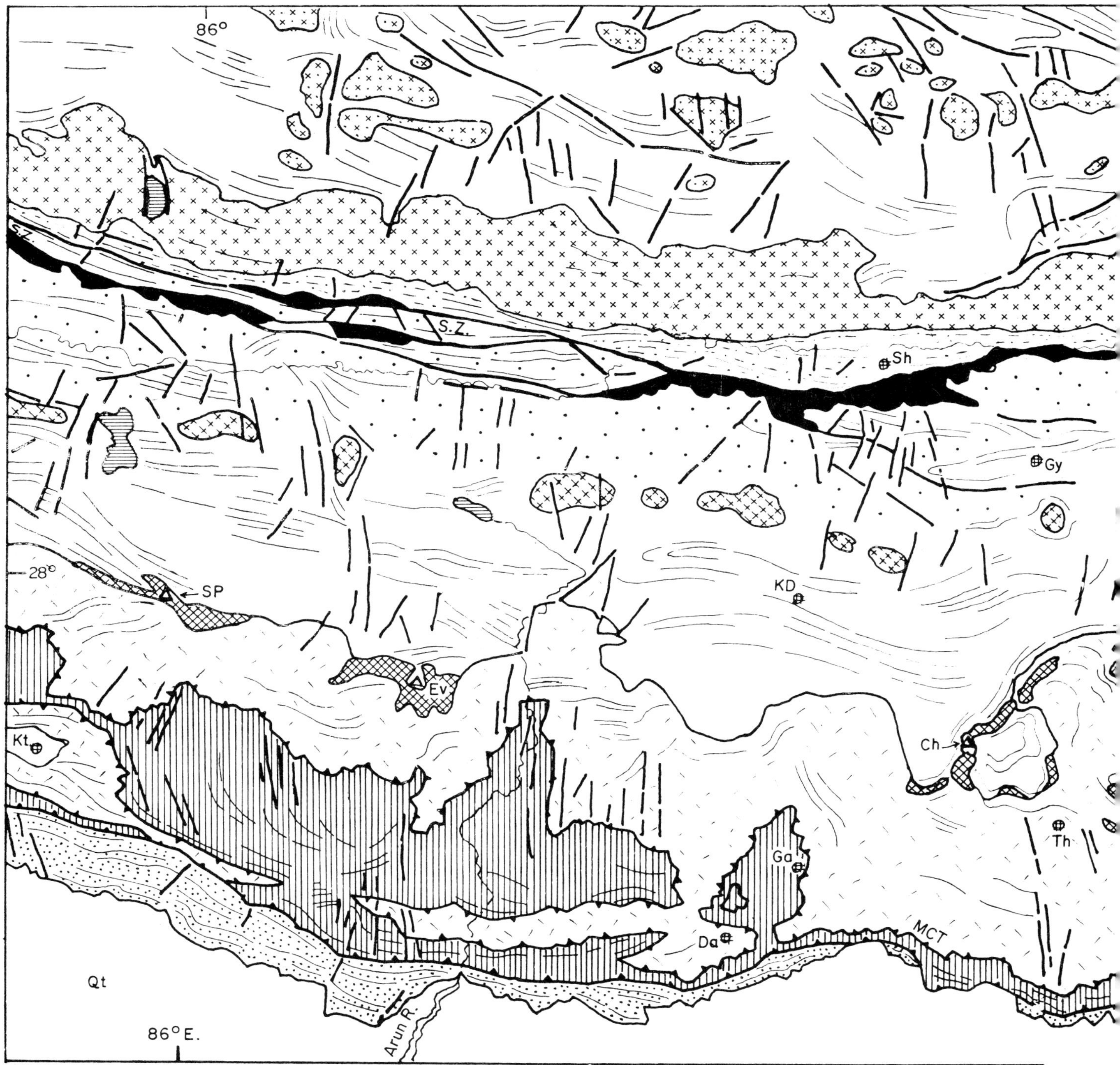

Fig. 136.

THE WIDER GEOLOGICAL FRAME OF THE BHUTAN HIMALAYA

Compiled form Gansser: Ophiolite belts Himalaya and Tibet, 1979; Geol. Map of Bhutan, 1981; and from Academia Sinica: Geol. Map Lhasa-Nyalam area 1979; as well as from the author's field observations.

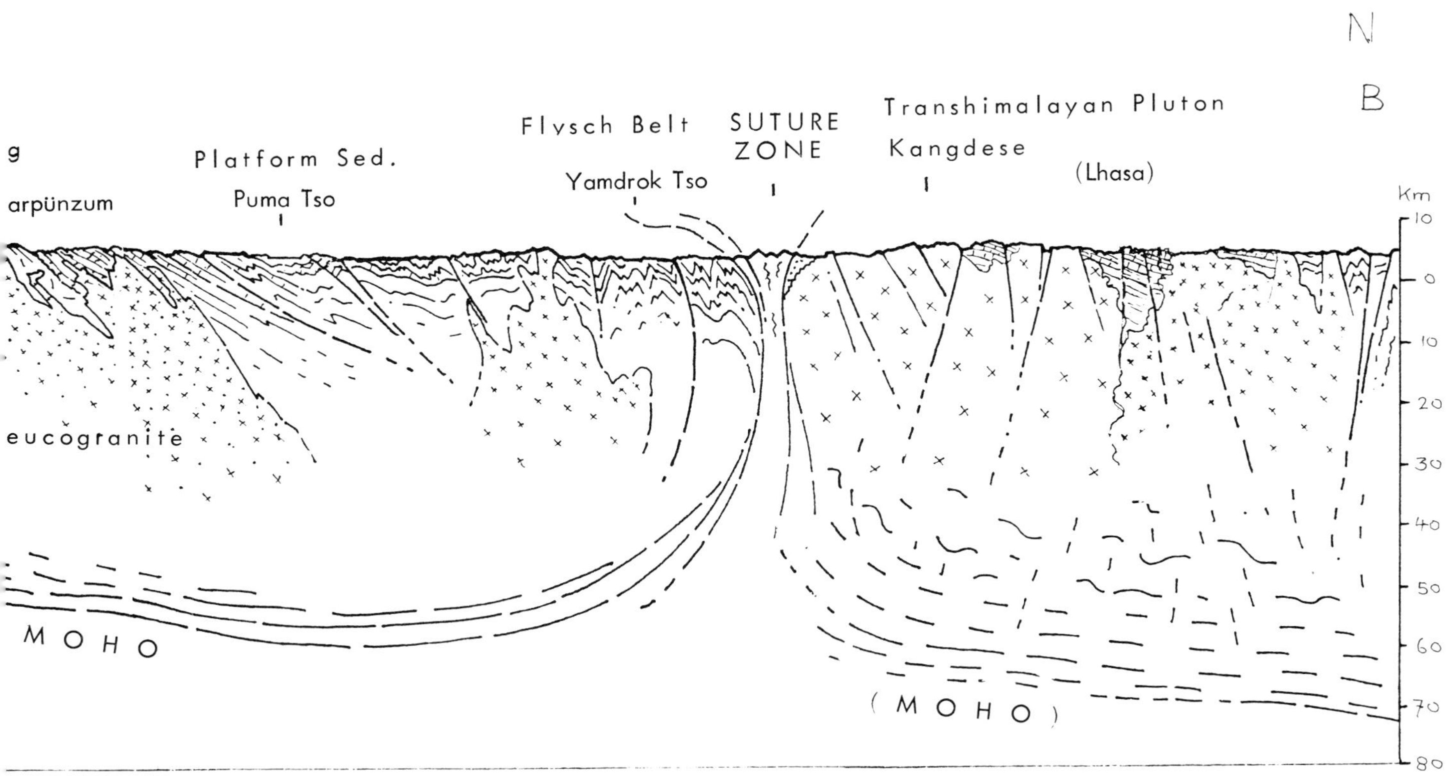
N
B
g
arpünzum
Platform Sed.
Puma Tso
Flysch Belt
Yamdrok Tso
SUTURE ZONE
Transhimalayan Pluton
Kangdese
(Lhasa)
Km
10
0
10
20
30
40
50
60
70
80
eucogranite
MOHO
(MOHO)

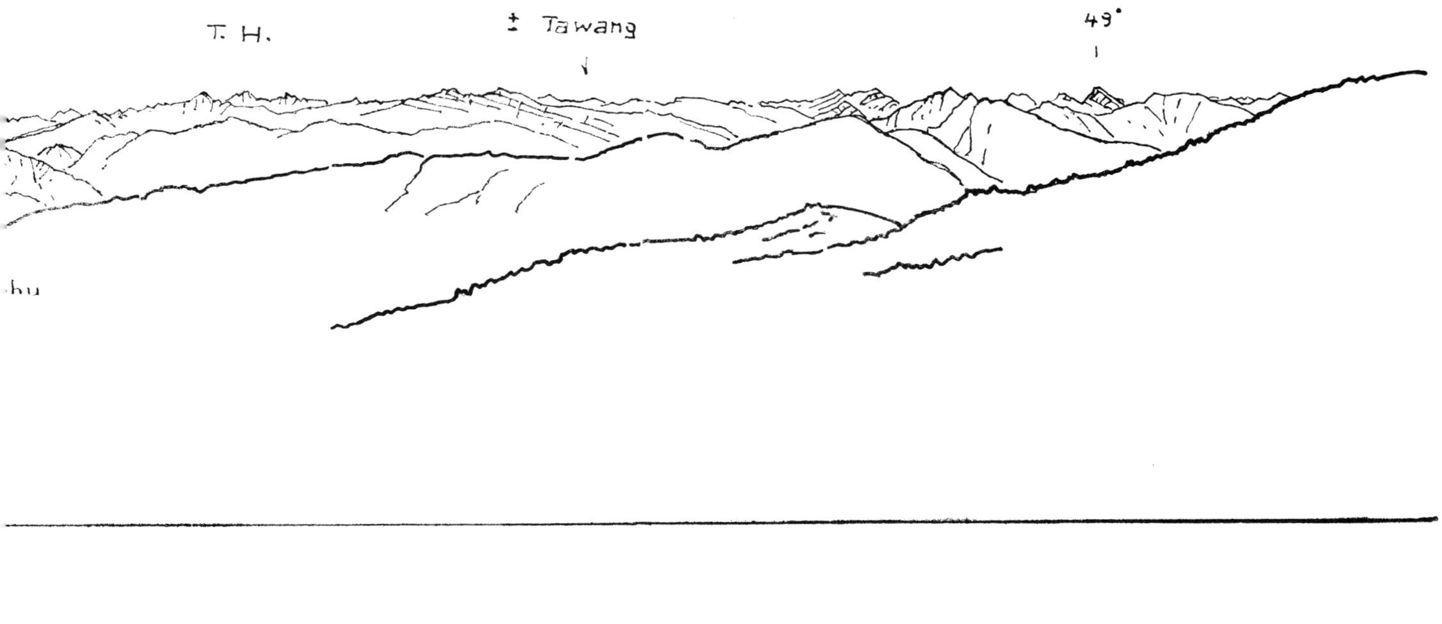
T. H.
Tawang
49°

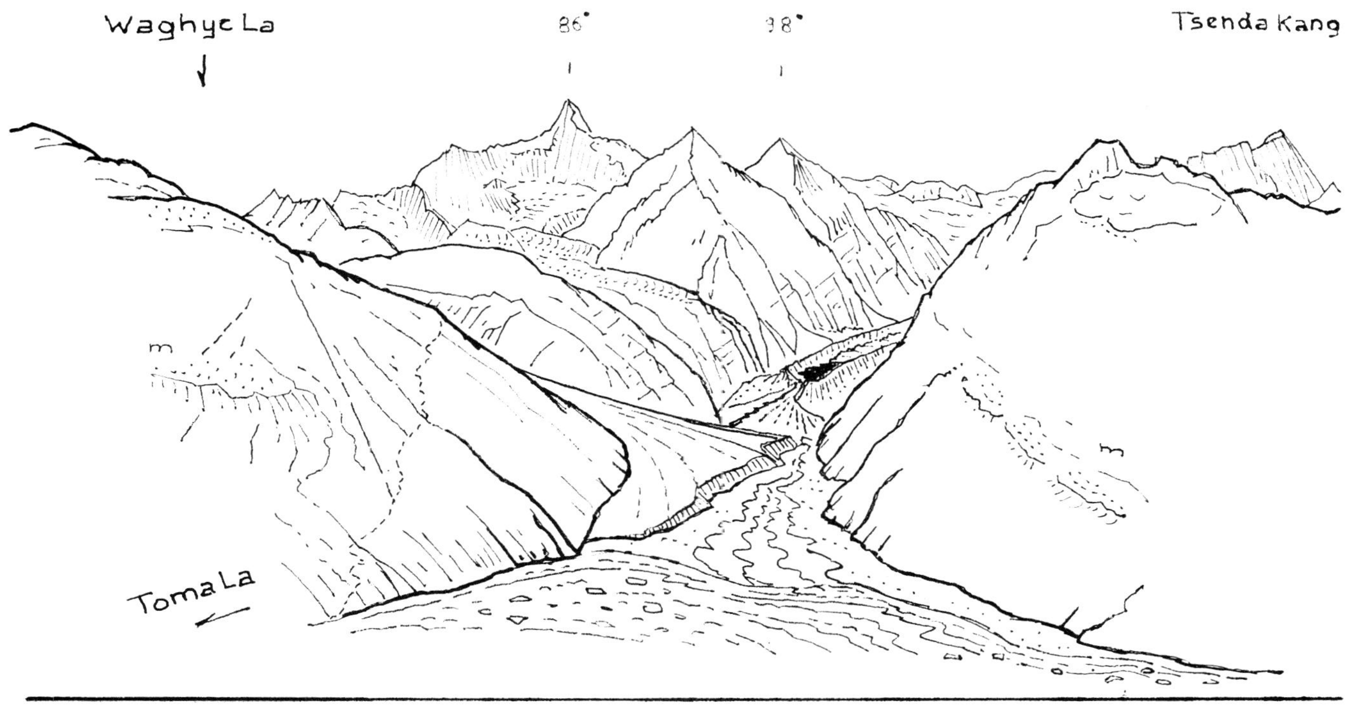

Fig. 139. The upper Pulumechu valley, view to E from Pt. 4270, E Masang Kang. Banded biotite migmatite mountains. Strong development of morains with old stage at m. Partly broken glacial lake at Tsenda Kang glacier. Has eroded the fan from Waghye La and subsequently the scree material is cut by the recent lake break from Masang Kang glacier valley foreground. (see also fig. 46).

the Indian shield by the Main Frontal Thrust, a recent steep thrust probably reaching into and displacing the Moho (Earthquake belt). 3. The Tibetan high plateau, beginning with the Transhimalayan Pluton (Kangdese belt) and separated from the Himalayan sector by the important Suture Zone.

In the Bhutanese Himalaya we recognize the Main Boundary Thrust (MBT), overriding a narrow, inconsistent Siwalik molasse belt and the Main Central Thrust (MCT) bringing the main crystalline, originally of Indian Shield origin, onto the Lower Himalayan metasediments. The latter represent a northern sedimentary cover of the Shield with Gondwana affinities.

The crystalline thrustmasses dominate the Bhutan Himalaya. They form rather flat thrust sheets with alternating basement type crystalline and highly metamorphosed metasediments. They have been thrown into large synforms and antiforms, with the mayor Masang Kang antiform limiting the crystalline belt to the N. The leucogranites, which frequently outcrop as laccolithic sills probably increase considerably at depth. They may have formed by palingenic melting at a depth of about 30 km, which represents the depth of the Indian Shield Moho. In northern Bhutan the Moho may be as deep as 60 km, increasing in depth from S to N. Direct information is not available, but we may assume that conditions are similar to those in other parts of the Himalaya. The Moho is probably less well defined and perhaps looses its sharp velocity changes when approaching the Suture Zone, where large slices of mantel material are involved.

The hade of the Suture Zone is shown steeply inclined to the S, at least in its upper part. This is based on the locations of earthquakes discussed by Chinese authors during the Tibet symposium. No sign of any Benioff Zone dipping under Tibet has been detected during the study of recent earthquakes. However, underthrusting may have been responsible for the large generation of the Transhimalayan (Kangdese) plutons between 110 and 40 my ago. The original picture changed drastically after the actual continental collision, thought to have happened during the Oligo-Miocene. It is therefore most difficult to draw a convincing picture of the present situation. No sharp Moho is expected below the Transhimalayan plutonic belt. A transitional zone must be assumed between a depth of 80 and 60 km. This deep transitional zone most likely continues below the high Tibetan plateau with an assumed crustal thickness of 70–80 km. This is further supported by the low traveltime of surface waves down to the base of the abnormally thick crust (*Chun* and *Yoshii*, 1977). The low velocities and the widespread Neogene to recent volcanism within Tibet (fig. 3) suggest a soft, hot, but thick crust, still being compressed in a NS direction but expanding to some extent in a EW direction. Its subrecent rise of 4000 m and its present rise by over 5 m/m a year is, considering the affected area of 2500000 km^2, the most fascinating geological event on our globe. The northern part of the Bhutan Himalaya is involved in this morphogenic phase, which resulted in the carving of some of the most beautiful and still practically unknown mountains of this fascinating country.

L. Selected Bibliography

Academia Sinica (1980): A Scientific Guidebook to South Xizang (Tibet) Organizing Committee, Symposium Tibet plateau pp 104.
- Beijing, China (with geol. map 1:150000).

Acharyya, S.K. (1970): Rangit Pebble Slate: A new formation from Darjeeling Foothills. - Ind. Minerals, 25, 1, p 61–64.
- (1974): Stratigraphy and Sedimentation of the Buxa Group, Eastern Himalaya. -Himalayan Geology, 4, p 102–116.
- (1975): Structure and Stratigraphy of the Darjeeling Frontal Zone, Eastern Himalaya. - Misc. Publ. Nr.24, Geol. Surv. India, p 71–90.

Allègre, C.J. and *Othman*, D.B. (1980): Nd-Sr isotopic relationship in granitoid rocks and continental crust development: a chemical approach to orogeneses. - Nature, 286, July, p 335–346.

Auden, J.B. (1934): Geology of the Krol belt. - Rec. Geol. Surv. India, 67, p 357–454.
- (1935): Traverses in the Himalaya. - Rec. geol. Surv. India, 69/2, p 123–167.
- (1970): Duscussion «On the Krol Nappe Hypothesis». - Journ. Geol. Soc. India, 11, 3, p 288–302.
- (1981): India's Former Crustal Neighbours. Proc. Indian natn. Sci. Acad., 47, A, 6, p 588–630.

Bhanot, V.B., *Singh*, V.P., *Kansal*, A.K. and *Thakur*, V.C. (1977): Early proterozoic Rb-Sr Whole-rock Age for Central Crystalline Gneiss of Higher Himalaya. Kumaon. - Jour. Geol. Soc. India, 18, 2, p 90–91.

Bhatia, S.B. and *Kanwar*, R.C. (Editors) (1975): Blaini and Related Formations. - Bull. Ind. Geol. Ass. (Special Issue) 8, 2, pp 279.

Bordet, P., *Cavet*, J. and *Pillet*, J. (1959): La faune de Phulchauki, près de Kathmandu (Himalaya du Nepal). - Bull. Soc. géol. France, 7, 2, p 3–14.

Chang, Cheng-Fa and Cheng, Hsi-ian (1973): Some tectonic features of the Mt. Jolmo Lumgma area, southern Tibet, China. - Sci. Sinica 16, p 257–265.

Chang, Chen-Fa, Zheng, Xi-lan, and *Pan, Yu-sheng* (1977): The Geological History, Tectonic Zonation and Origin of Uplifting of the Himalayas. - Institute of Geology, Academia Sinica, Peking, pp 17.

Chappel, B.W. and *White, A.J.R.* (1974): Two contrasting granite types. - Pacif. Geol. 8, p 173–174.

Chun, K.Y. and *Yoshii*, T. (1977): Crustal structure of Tibetan Plateau: A surface wave study by a moving window analysis. - Bull. Seimol. Soc. Am., 67, p 735–750.

Das, A.K., *Bakliwal*, P.C. and *Dhoundial*, D.P. (1975): A brief outline of the Geology of parts of Kameng District (NEFA). - Geol. Surv. India, Misc. Publ. Nr. 24 (1), p 116–127.

Frank, W., *Hoinkes*, G., *Miller*, C., *Purtscheller*, F., *Richter*, W. and *Thöni*, M. (1973): Relations between Metamorphism and Orogeny in a Typical Section of the Indian Himalayas. -Tscherm. mineral. petrogr. Mittl., 20, p 303–332.

Frank, W., *Thöni*, M., *Purtscheller*, F. (1977): Geology and Petrography of Kulu-South Lahul Area. - Colloq. intern. C.N.R.S., 268, Ecologie et Géologie de l'Himalaya, p 147–172.

Fuchs, G. and *Sinha*, A.K. (1978): The Tectonics of the Garhwal-Kumaun, Lesser Himalaya. - Jahrb. Geol. B.-A., 121,(2) p 219–241.

Gansser, A. (1964): Geology of the Himalayas, - Wiley-Interscience, London, p 289.
- (1966): Geological Research in the Bhutan Himalaya. - The Mountian World, Swiss Foundation of Alp. Res. Zürich, p 88–97.
- (1970): Lunana, the Peaks, Glaciers and Lakes of Northern Bhutan. - The Mountain World, Swiss Foundation of Alp. Res. Zürich, p 117–131.
- (1974): The Himalayan Tethys.-Riv. Ital. Paleont. Strat., Mem. XIV, p 393–411.
- (1977): The great suture zone between Himalaya and Tibet - a preliminary account. - Colloq. intern. C.N.R.S., 268, Ecologie et Géologie de l'Himalaya, pp 181–191.
- (1978): Proposals for an investigation of dangerous glacial lakes in North Bhutan. - Report for the Gov. of Bhutan. (not published)
- (1979): Ophiolithic belts of the Himalayan and Tibetan Region, Sheet 4 of International Atlas of Ophiolites with explanatory notes. IGCP, Project 39. - Geol. Soc. Am., Map and Chart Series, MC-33.
- (1980): The Peri-Indian suture zone. - Mémoire du B.R.G.M., 115, p 140–148.

Gansser, U., *Gansser*, A. and *Olschak*, B.C. (1971): Bhutan-Land of Hidden Tresures. - Georg Allen and Unwin Ltd. London, pp 144.

Geological Survey of India (1971): Recent Geological Studies in the Himalaya, Abstracts.

Ghose, N.C. and *Sinkgh*, N.K. (1977): Experimental study of granitic rocks of Darjeeling (West Bengal, India) and its application to the origin of Himalayan granites. - Tectonophysics, 43, p 23–40.

Godwin-Austen, H.H. (1868): Note on geological structure of the country near foot of Hills in the Western Bhootan Dooars - J. As. Soc. Beng. 37, 1, p 1–27.

Gosh, A.M.N. (1952): A new coalfield in the Sikkim Himalaya. - Curr. Sci. 21, 7, p 179–180.

Gupta, V.J. (1971): Devonian fossils from the Lesser Himalayan Zone of Central Bhutan. - Curr. Sci., 40, 18, pp 490.

Gupta, V.J. and Waterhouse, J.B. (1979): Permian Fossils from Sikkim and Bhutan. - Bull. Ind. Geol. Assoc. 12, 2, p 253–255.

Hamet, J. and *Allegre*, J.-C. (1976): Rb-Sr systematics in granite from central Nepal (Manaslu): significance of the Oligocene age and high $^{87}Sr/^{86}Sr$ ratio in Himalayan orogeny. - Geology 4, p 470–472.

Hänny, R. (1966): Bhutan, unbekannter Himalaya. - «Die Alpen», Quartalsheft 3, pp 10.

Hayden, H.H. (1907): The geology of the provinces of Tsang and U in Tibet. - Geol. Survey India Mem., 36, 2, p 122–201.

Heim, A. (1938: The Himalayan border compared with the Alps. Rec. geol. Surv. India, 72, p. 413–421.

Heim, A. und *Gansser*, A. (1939): Central Himalaya. - Geological Observations of the Swiss Expedition 1936, Denkschr. der Schweiz. Naturf. Ges. 73, 1, p 247.

Honegger, K., *Dietrich*, V., *Frank*, W., *Gansser*, A., *Thöni*, M. und *Trommsdorff*, V. (1982): magmatism and metamorphism in the Ladakh Himalayas. - Earth planet. Sci. Lett. (in press).

Hsu, J. (1978): On the Paleobotanical Evidence for Continental Drift and the Himalyan Uplift. - The Paleobotanist, Vol. 25, p 131-142.

Itihara, M., *Shibasaki*, T. and *Miyamoto*, N. (1972): Photogeological survey of the Siwalik Ranges and the Terai Plain, Southeastern Nepal. - Jour. of Geosc. Osaka City Univ., 15, 4, p 77–96.

Jangpangi, B.S. (1974): Stratigraphy and Tectonics of parts of Eastern Bhutan. - Himalayan Geology 4, p 117-136.
- (1976): Thermal Springs of Southern Bhutan. - Geol. Surv. India, Misc. Publ. Nr. 24, 11, p 460-463.
- (1978): Stratigraphy and Structure of Bhutan Himalaya. - In Tectonic Geology of the Himalaya (Saklani Ed.), p 221–242.

Jangpangi, B.S., *Pushikar Singh, Guhar Sarkar,* T.K. and *Jagannatha,* B.R. (1975): Stratigraphy and Tectonics of the Precambrian Formations in Bhutan. - Rec. Geol. Surv. India, Vol 106, 2, p 86–95.

Karan, P.P. (1967): Bhutan, a Physical and Cultural Geography. - University of Kentucky Press, Lexington, pp 103.

Krishnan, M.S. (1960): Geology of India and Burma. - Higginbothmas, Madras, pp 604.

Künsel, a weekly official Bulletin of the Royal Gevernment of Bhutan (1978); XIII, Nr. 10 and 12.

Lahiri, A., (1941): Geology of the Buxa Duars. - Quart. J.geol. Soc. India, 13, (1) p 1–62.

Le Fort, P. (1973): Les Leucogranites de l'Himalaya, sur l'exemple du granite du Manaslu (Népal central). - Soc. Geol. France. Bull., 7, pp 555-561.

- (1975): Himalayas: The collided range. - Present knowledge of the continental arc. Amer. J. Sci. 275-A, p 1–44.

- (1981): Manaslu Leucogranite: A Collision Signature of the Himalaya. A Model for its Genesis and Emplacement. Journal of Geoph. Res. 86, B11, 10545–10568.

Le Fort, P. et *Pêcher,* A. (1974): Les gneiss oeillés du moyen pays népalais: un ensemble volcano-sedimentaire acide d'âge paléozoic ou plus ancien en Himalaya. - C.R. Acad. Sc. Paris, 278, p 3283–3286.

Lexique Stratigraphique International, Vol III, Asie, Fasc. 8 (1957), pp 404.

Li Chi-chun and *Cheng Pen-hsing* (1980): Recent research on glaciers on the Chinghai-Tibet Plateau. - Iahs-Aish Publ. Nr. 126, p 121–127.

Liao, Zhijie (1981): Hot Springs in Tibet. Peking University, Geol. Dept. Scientific Book House. (Peking) pp 170.

Lombard, A. (1958): Un itinéraire géologique dans l'Est du Népal (Massif du Mont Everest). Mém. Soc. Helv. Sci. nat., 82, p 107.

Mallet, F.R. (1875): On the geology and mineral resources of the Darjiling District and the Western Duars. - Mem. seol. Surv. India, 11, p 1–50.

Mehra, G.N. (1974): Bhutan, Land of the Peaceful Dragon. - Vikas publishing House Ltd., Delhi, pp 151.

Misch, P (1936): Einiges zur Metamorphose des Nanga-Parbat. - Geol. Rdsch. 27, p 79–81.

Mu An-Tze, Wen Shih-Hsuan, Wang Yi-Kang, Chang Ping-Kao and *Yin Chi-Hsiang* (1973): Stratigraphy of the Mount Jolmo Lungma region in southern Tibet, China. - Scientia Sinica, 16, (1) p 96–111.

Müller, F. (1959): Eight months of glacier and soil researche in the Everest region. - In The Mountain World, Swiss Foundation of Alpine Researche, Zürich, p 191–208.

- (1980): Present and late Pleistocene equilibrium line altitudes in the Mt. Everest region - an application of the glacier inventory. - Iahs-Aish, Publ. Nr. 126, p 75–94.

Nakamura, N. (1974): Determination of REE, Ba, Fe Na and K in carbonaceous and ordinary chondrites. - Geochim. et Cosmochim. Acta, 38, p 757–775.

Nakata, T. (1972): Geomorphic History and Crustal Movements of the Foot-Hills of the Himalayas. - Sc. Rep. Tohoku Univ., 7th Ser. 22, 1, p 39–177.

Nautiyal, S.P., *Jangpangi,* B.S., *Singh,* P., *Sarkar,* T.K.G., *Bhate,* V.D., *Raghavan,* M.R. and *Sahai,* T.N. (1964): A preliminary note on the geology of Bhutan Himalaya. - Int. Geol. Congr. 22nd Sess, 11, p 1–14.

Nisbet, E.G., *Dietrich,* V.J. and *Esenwein,* A. (1979): Routine trace element determination in silicate minerals and rocks by X-ray fluorescence. - Fortschr. Miner. 57, 2, p 264–279.

Olschak, B.C. (1971): Bhutan, das Land der unbestiegenen Gletscher als neues UNO-Mitglied. - «Die Alpen», Quartalsheft 4, pp 7.

- (1979): Ancient Bhutan, a study of Early Buddhism in the Himalayas. - Swiss Foundation of Alp. Res., Zürich, pp 222.

O'Rourke, J.E. (1962): The Stratigraphy of Himalayan Iron Ores - Amer. J. Sci 260, p 294–302.

Pantic, N., *Hochuli,* P.A. and *Gansser,* A. (1981): Jurassic palynomorphs below the Main Central Thrust of East Bhutan (Himalayas). Eclogae geol. Helv. 74, 3, 883–892.

Pilgrim, G.E. (1906): Notes on the geology of a portion of Bhutan. - Rec. geol. Surv. India, 34, (1) p 22–30.

Pilgrim, G.E. and *West,* W.D. (1928): The structure and correlation of the Simla rocks. - Mem. geol. Surv. India, 53, pp 140.

Poulose, K.V. (1975): Tourmaline-Granites and Pegmatites associated with the Thimphu and the Chekha of Bhutan Himalaya. - Geol. Surv. India, Misc. Publ. Nr. 24, (1) p 257–262.

Sengupta, S. and *Raina,* P.L. (1978): Geology of parts of Bhutan Foothills adjacent to Darjeeling District. - Indian Journ. of Earth Sc. 5, (1) p 20–33.

Shand, S.J. (1950): Eruptive Rocks 4th Ed.). - John Wiley, New York.

Singh, P. (1973): A note on the Fossiliferous Formations in Lesser Himalaya of Nepal and Bhutan. - Himalayan Geology, 3, p 372–380.

Stöcklin, J. (1977): Structural correlation of the Alpin ranges between Iran and Central Asia. - Mém. h. sér. Soc. géol. France, 8, p 333–353.

- (1980): Geology of Nepal and its regional frame. - J. geol. Soc. London, 137, p 367–373.

Stöcklin, J., *Termier,* G. and *Bhattarai,* K.D. (1977): A propos de roches fossilifères de Chandragiri (Mahabharat Himalaya du Nepal). - Bull. Soc. géol. France, 19, p 367–373.

Survey of India (1972): 1:250000 topographical map of Bhutan. Printed at the 101 (H.L.O.) Printing Group of Survey of India.

Termier, G. and *Gansser,* A. (1974): Les séries dévoniennes du Tang Chu (Himalaya du Bhoutan). - Eclogae geol. Helv. 67, (3), p 587–596.

Tuttle, O.F. and *Bowen,* N.L. (1958): Origin of granite in the light of experimental studies in the system Na Al Si_3O_8 - KAl Si_3O_8 - SiO_2-H_2O. - Geol. Soc. Am. Mem., 74, p 1–153.

Tyson, J. (1978): Exploring Bhutan's North-west. - The Alpine Journal 83, p 183–190.

Valdiya, K.S. (1969): Stromatolites of the Lesser Himalayan carbonate formations and the Vindhyans. - J. geol. Soc. India, 10, p 1–25.

Verma, P.K. and *Tandon, S.K.* (1976): Geologic Observations in a part of the Kameng District, Arunachal Pradesh (NEFA). - Himalayan Geology, 6, p 259–286.

Wadia, D.N. (1957): Geology of India. - 3rd Edn., Macmillan London, pp 636.

Ward, M. (1966): Some Geographical and Medical Observations in North Bhutan. - The Geogr. Journ., 132, (4) p 493–506.

Winkler, M.G.F. and *Ghose,* N.C. (1973): Further data on the eutectics in the system Oz-Or-An-H_2O. - Neues Jahrbuch. Min. Monatschr., 11, p 481–484.

Yang, L. and *Liu,* T. (1974): Neotectonic movements of Mt. Qomolangma District. - Scientia Geol. Sinica., 3, p 209–220.

Subject Index

Abstract

The Bhutan Himalaya, one of the least known sections of the 2500 km long mountain range has been investigated geologically during 5 expeditions between 1963 and 1977. Previous work is scarce and only after 1960 began the foothill-mapping of the Geological Survey of India. The Bhutan Himalaya can be subdivided into the geological units well known for the whole range:

The Subhimalaya with the Siwalik sediments. The *Lower Himalaya* with the Late-Precambrian metasediments and marginal remnants of Gondwana rocks along the Main Boundary Thrust. The crystalline units of the *High Himalaya* which cover with over 15 km thick thrust sheets the greater part of the country. They are transgressed by the unmetamorphic, fossiliferous sediments of the *Tethyan Himalaya* preserved in the Lingshi basin in the NW and the Tang Chu basin in the center. They also just cover the highest crystalline rocks along the Tibetan border from where they merge with the main Tibetan Tethyan rocks. The *Suture Zone* follows to the N of Bhutan along the Tsangpo river.

The Siwaliks are locally missing by erosion and subsequent relief thrusting of the Lower Himalayan units. These Siwalik gaps are only known from the Bhutanese foothills and may coincide with the northwards directed crystalline spur of the Shillong shield which narrows the voreland to only 35 km. The age of the sediments of the Lower Himalaya is still disputed. They may represent as late Precambrian Daling schists the northern cover of the Indian Shield, together with typical Gondwana relics. In spite of the complicated tectonics a reversed metamorphism can be recognized, increasing towards the Main Central Thrust, which is not every where well recognizable. On the other hand, along its southern front in eastern Bhutan occurs a thin tectonic wedge of calcschists with an Oxfordian florula.

The crystalline thrust sheet, with widespread granitoid gneisses and migmatites contains intercalated metasediments rich in high metamorphic marbles and calc-silicates. They are well displayed in the sout wall of Chomolhari mountain. In this northern area, with the highest metamorphic grades and indications of a still little understood reversed metamorphism, discordant intrusions of leucogranites are widespread. Their extremly high Sr isotope ratio suggests melting of an acide crystalline crust. The leucogranites cut through all structures and postdate the regional Himalayan metamorphism. The latter phase has recrystallized all previously already metamorphosed rocks but many relics of a pre-Himalayan metamorphism and structures are still preserved. Some granites, developing from migmatites in the N (Masang Kang region) gave a preliminary isochrone of about 400 my. The frequent NS directed folds and lineations, together with a marked young fracture pattern may reflect old but reactivated NS structures of the northern Indian Shield.

Little known is the morphology and glaciology of Bhutan. During the new investigations five older glacial stages were observed, the lowest relicts at about 3000 m above sea level. Best developed is the 100 years old Thanza stage averaging 4100 m while the recent glaciers end between 4200 and 4500 m, about 600 m lower than in the dryer Everest region. A marked peneplane was formed during the Thanza stage, culminating with about 5000 m over central Bhutan which may coincide with a regional cross high outlined by the Black Mountains further to the south.

Zusammenfassung

Der Bhutan Himalaya, eines der am wenigsten bekannten Teilstücke der 2500 km langen Gebirgskette wurde zwischen 1963 und 1977 durch 5 Expeditionen geologisch untersucht. Vorhergehende Arbeiten sind selten und erst nach 1960 hat das Geological Survey von Indien mit der Kartierung der Fusshügelzonen begonnen. Die für den Gesamthimalaya bekannte regionale Einteilung kann auch für den Bhutan Himalaya angewandt werden: Der *Subhimalaya* mit den Siwalik Sedimenten. Der *Niedere Himalaya* mit den spätprekambrischen Meta-Sedimenten und den randlichen Einschaltungen von Gondwana Gesteinen längs des «Main Boundary Thrust». Die kristallinen Einheiten des *Hohen Himalaya* welche mit den über 15 km mächtigen Schubmassen den grösseren Teil des Landes bedecken. Unmetamorphe, fossilführende Sedimente des *Tethys Himalaya* sind im Lingshi Becken im NW und im Tang Chu Becken im Zentrum erhalten. Sie bedecken ebenfalls die höchsten kristallinen Einheiten längs der tibetanischen Grenze und verbinden sich mit der Hauptzone der tibetanischen Tethys Ablagerungen. Die eigentliche *Suture Zone* folgt nördlich von Bhutan längs des Tsangpo Flusses.

Die Siwalikketten sind durch Erosion und darauf folgende Relief-Überschiebung des Niederen Himalayas local unterbrochen. Diese Unterbrüche sind nur längs des Bhutan Himalaya bekannt und dürften mit dem nördlichen kristallinen Sporn des Shillong Schildes zusammenhängen, wobei das Vorland auf nur 35 km eingeengt wurde. Das Alter der Sedimente des Niederen Himalayas ist immer noch sehr umstritten. Sie dürften als spätprekambrische Daling Schiefer der nördlichen Sedimentbedeckung des Indischen Schildes entsprechen, zusammen mit den typischen reliktischen Gondwana Gesteinen. Trotz der komplizierten Tektonik kann eine umgekehrte Metamorphose festgestellt werden, die gegen den «Main Central Thrust» hin zunimmt. Dieser ist jedoch nicht immer leicht erkennbar. Anderseits ist längs seiner südlichen Front im östlichen Bhutan eine tektonische Schuppe von Kalkschiefern eingeschaltet mit einer Flora von Oxford Alter.

Die kristalline Hauptdecke zeigt Aufschlüsse von eingeschalteten Meta-Sedimenten in Form von hochmetamorphen Marmoren und Kalksilikatlagen. Diese sind in den Südwänden des Chomolharigebirges besonder gut aufgeschlossen. Hier im Norden, innerhalb der höchstmetamorphen Gesteine mit einer noch wenig verstandenen Umkehr des Metamorphosengrades finden sich diskordante Intrusionen von Leucograniten. Das aussergewöhnlich hohe Sr-Isotopenverhältnis deutet auf Aufschmelzung von sauren Krustengesteinen hin. Die Leukogranite durchbrechen sämtliche Strukturen und sind auch jünger als der regionale Himalaya-Metamorphismus. Letzterer hat sämtliche älteren metamorphen Gesteine überprägt, trotzdem sind noch Relikte von metamorphen und strukturellen Vorhimalaya-Phasen vorhanden. Einige Granite, die sich im Norden (Masan Kang Region) aus Migamtiten entwickelt haben, gaben eine vorläufige Isochrone von ungefähr 400 my. Die häufigen NS gerichteten Faltungen und Lineationen, zusammen mit ausgeprägten jüngsten Bruchsystemen, dürften einer alten aber reaktivierten NS Richtungen des nördlichen Indischen Schildes entsprechen.

Wenig ist über die Morphologie und Glaziologie Bhutans bekannt. Während der neueren Untersuchungen konnten fünf ältere Glazialstadien festgestellt werden, mit den ältesten Relikten bei ungefähr 3000 m Meereshöhe. Am besten entwickelt ist das 100 Jahre alte Thanza Stadium bei ungefähr 4100 m während die rezenten Gletscher zwischen 4200 und 4500 m enden, 600 m tiefer als in dem trockeneren Everest Klima. Während des Thanza Stadiums hat sich eine ausgeprägte Peneplane gebildet, die bei etwa 5000 m über Zentralbhutan kulminiert, und mit einem durch die Black Mountains weiter südlich ausgeprägten Querhoch zusammenhängen dürfte.

Resumé

L'Himalaya du Bhoutan, une des sections les moins bien connues de la chaîne entière, 2500 km de long, a été étudié entre 1963 et 1977 par 5 expéditions géologiques. Les recherches précédantes était peu nombreuses et ce n'est qu'après 1960 que le Service Géologique de l'Inde se mettait a cartographier les collines du Bas Himalaya. La subdivision régionale de la chaîne Himalayenne peut aussi être utilisée pour l'Himalaya du Bhoutan.: Le *Sub-Himalaya* avec les sédiments des Siwaliks. Le *Bas-Himalaya* avec les meta-sédiments Précambriens tardifs et les intercalations bordières des Gondwanas le long du «Main Boundary Thrust». Les unités cristallines du *Haut-Himalaya* dont les masses allochtones épaisses de plus de 15 km, recouvrent la plus grande partie du pays. Les unités non-métamorphiques et fossilifères du *Tethys Himalaya* sont préservées dans les bassins de Lingshi au NW et Tang Chu au centre. Elles recouvrent aussi les chaînes cristallines les plus hautes le long de la frontière avec le Tibet ou ils se connectent au dépots tethienne du Tibet. La *Zone de la Suture* suit le fleuve Tsangpo au N du Bhoutan.

Les chaînes des Siwaliks sont localement coupées par l'érosion et les lacunes sont recouvertes par les charriages du Bas-Himalaya. (Charriage au relief). Ces lacunes, connues seulement le long du Sub-Himalaya Bhoutanais paraissent en relation avec l'extension vers le N des roches cristallines du bouclier de Shillong, réduisant l'avant pays a 35 km de largeur. L'âge des sédiments du Bas-Himalaya est encore en discussion. Les schistes du type Daling, probablement Précambrien tardif représenteraient la continuation septentrionale de la couverture sédimentaire du bouclier Indien, inclus les roches typiques du Gondwana. Malgré la tectonique fort compliquée on reconnaît que le métamorphism croît du bas vers le haut, et augmente vers le «Main Central Thrust». Cette zone de chevauchement n'est pas toujours bien exposée. D'autre part, le long de son front méridional au Bhoutan oriental on reconnaît une écaille tectonique de calcschistes avec des flores d'un âge Oxfordien.

La grande nappe cristalline contient des zones sédimentaires fort métamorphisées en forme de différents types de cipollins et des gneisses calcitiques. Ils sont bien visible dans l'imposante parois sud du Chomolhari. Dans ce secteur au N du pays, caracterisé par un très haut degré de métamorphism, localement à l'envers (un fait toujours difficile à expliquer) nous trouvons de remarquables intrusions de leucogranites discordantes. La relation extrèmement haute des isotopes Sr des granites indique la fusion d'une croute continentale acide. Les leucogranites coupent toutes les structures et ils sont plus jeunes que le métamorphism régional de la phase Himalayenne. Cette phase a recristallisée toute la roche métamorphisée antérieurement mais tout de même des restes d'un métamorphism et de structures pré-Himalayennes sont encore préservés. Au N des granites naissant des migmatites (region du Masang Kang) ont donnés une isochrone préliminaire de 400 ma. Les plis et les linéations souvent alignées NS ainsi que des jeunes systèmes de fractures semblent avoir repris la direction des anciennes structures septentrionales du bouclier Indien.

L'histoire glaciologique et la morphologie du Bhoutan sont très mal connues. Les nouvelles investigations ont démontrées cinq périodes glaciales, dont les restes les plus bas se trouvent à environ 3000 m audessus du niveau de la mer. La période de Thanza, bien developpée, a été formée il y a 100 ans à une altitude d'environ 4100 m. Les glaciers récents cependant descendent à des niveaux situés entre 4200 et 4500 m, c'est-à-dire 600 m plus bas que les glaciers de l'Everest, situés dans un climat plus sec. Une pénéplaine bien marquée a été formée pendant la période de Thanza. Elle culmine avec 5000 m dans le Bhoutan central, un phénomène qu'on pourrait corréler avec le haut transversal des Black Mountains situé plus au S.

Riassunto

L'Himalaya del Bhutan, una delle zone meno conosciute della catena di montagne lunga 2500 km, è stato studiato tra gli anni 1963 e 1977 da cinque spedizioni geologiche. I lavori precendenti sono pochi, e solo dopo l'anno 1960 ha cominciato la cartografia nelle colline meridionali dal servizio geologico dell'India. Nel Himalaya del Bhutan riconosciamo le suddivisioni regionali proprie a tutta la catena: Il *Sub-Himalaya* con i sedimenti del Siwalik. *Il Basso Himalaya* con i meta-sedimenti tardo-precambriani e resti marginali di roccie del Gondwana lungo il «Main Boundary Thrust». Le unità cristalline del *Alto Himalaya* le quali, con un espessore di oltre 15 km ricoprono una gran parte del paese. Sono ricoperti da una transgressione di sedimenti non metamorfici, fossiliferi del *Tethys Himalaya,* che si trovano preservati nel bacino di Lingshi al NW come anche in quello del Tang Chu nel centro. Ricoprono poi appena le parti piu alte delle catene cristalline lungo la frontiera con il Tibet dove si collegano con i depositi della Tethys Tibetana. La *Zona della Sutura* segue più al nord del Bhutan, lungo il fiume Tsangpo.

A causa di erosione e susseguente ricoprimento tettonico del Basso Himalaya una parte della formazione Siwaliks manca. Queste interruzioni sono conosciute solamente lungo il Sub-Himalaya del Bhutan e potrebbero essere in relazioni con uno sperone diretto verso il N delle roccie cristalline dello scudo di Shillong. Con questo l'avanpaese si riduce a solo 35 km di larghezza. L'età dei sedimenti del Basso Himalaya è sempre ancora discussa. Potrebbero corrispodere con gli scisti precambrici di Daling alla normale copertura N dello scudo Indiano, assieme a dei relitti delle roccie Gondwana. Malgrado la forte tettonizzazione un metamorfismo inverso è rinoscibile, aumentando verso il «Main Central Thrust», il quale però non è sempre ben esposto. D'altra parte, al suo bordo meridionale nel Bhutan del est si trova una lamella tettonica di calcescisti con una flora del Oxfordiano.

La grande falda cristallina, ricca in gneiss granitoidi e migmatiti contiene intercalazioni di metasedimenti in forma di marmi e calcefiri di un alto grado metamorfico. Sono assai ben visibili nella parete sud della montagna Chomolhari. In questa zone nord dove notiamo il più alto grado di metamorfismo, localmente inverso, – un fatto poco compreso –, troviamo estese intrusioni discordanti di leucograniti. L'altissima proporzione di isotopi Sr suggerisce fusione di una crosta acida cristallina. I leucograniti tagliano tutte le strutture e sono posteriori al metamorfismo regionale della fase Himalayana. Quest'ultima fase à recristallizzato tutte le roccie già metamorfosate, aunque tanti relitti d'un metamorfismo e di strutture pre-Himalayane sono ancora preservati. Alcuni graniti, formati da migmatiti nella parte nord (Masang Kang) hanno dato un'isochrona preliminare di ca. 400 ma. Le frequenti piegature e lineazioni in direzione NS come anche un systema di fratture d'étà tardive protrebbero suggerire una riattivazione di vecchie strutture NS della parte nord dello scudo Indiano.

Poco conosciuta è la morfologia e la glaziologia del Bhutan. Durante le nuove investigazioni 5 stadi glaciali sono stati osservati, con i relitti più bassi a ca. 3000 m sopra il livello del mare. Ben preservato è lo stadio di Thanza, di ca. 100 anni fà, a un promedio di 4100 m, mentre i ghiacciai recenti terminano tra i 4200 e 4500 m, approssimativamente 600 m più bassi che nella regione più arida del Everest. Durante lo stadio di Thanza fù formato un altiplano, culminando a 5000 m sopra il Bhutan centrale e che potrebbe coincidere con un'elevazione transversale marcata per i Black Mountains situati più al sud.